Sampling for the Measurement of Odours

Scientific and Technical Report No.17

Sampling for the Measurement of Odours

P. Gostelow, P. Longhurst, S. A. Parsons and R. M. Stuetz

IWA Publishing

Published by IWA Publishing, Alliance House, 12 Caxton Street, London SW1H 0QS, UK
Telephone: +44 (0) 20 7654 5500; Fax: +44 (0) 20 7654 5555; Email: publications@iwap.co.uk
Web: www.iwapublishing.com

First published 2003
Reprinted 2007

Edited and typeset by Steve Turrington, London, UK.
Index prepared by Indexing Specialists, Hove, UK.

British Library Cataloguing in Publication Data
A CIP catalogue record for this book is available from the British Library

Library of Congress Cataloging- in-Publication Data
A catalog record for this book is available from the Library of Congress

ISBN: 1 84339 033 7
ISBN 13: 9781843390336

Contents

Preface

Nuisance caused by odours is a major problem for waste management industries, requiring assessment, abatement and control. To control an odour nuisance, it must first be quantified. This is difficult for odours as there is no single parameter which can easily describe odour nuisance, and a range of measurements may be required. Many odour measurements cannot be performed on site, meaning that samples must be preserved for laboratory analysis.

For odours to cause nuisance, they must be transported from source to receptor. Although odours may be measured at receptor locations, for assessment and control purposes it is more useful to measure at source. This is because little dispersion will have occurred, so that fewer measurements are required and concentrations will be higher. In addition, there is less opportunity for odours from other sources to interfere with measurements. This report focuses on source measurements to determine the current or potential odour impact of any process.

For source measurements to be useful, it is not sufficient to simply measure concentrations. It is the rate of odour emission that is of paramount importance, if odour sources are to be suitably assessed and controlled. Emission rate measurements provide a means of ranking odour sources, of assessing their impact on receptors through dispersion modelling, and are essential design inputs for odour control schemes.

The original aim of this study was to review European and other internationally available research studies to learn best practice for odour source sampling, sample collection and preservation, and odour measurement in waste management applications. The review shows that there are a great variety of techniques available, particularly for source sampling. Unfortunately, many of these techniques can interfere with the odour sources or otherwise fail to adequately represent odour emissions. This has implications for the quality of samples obtained and their usefulness for assessing odour impact. Although some techniques are clearly better than others for various waste management applications, this report emphasises interferences and sources of error that can be introduced. It highlights these difficulties to make the practitioner aware of their implications, indicates established practice and highlights further research needs.

For odour measurement, the situation is clearer due to developments in standardisation, which have reduced much of the variation in results and helped to identify best practice in this area. Unfortunately, many difficulties still remain: in particular the lack of sensitivity of some methods, and relationships between different odour measurement techniques.

The need for this report was identified by the Engineering and Physical Sciences Research Council (EPSRC) Engineering Network for the Control and Prevention of Odour Emissions, a UK discussion platform that promotes multidisciplinary research in the areas of odour measurement, modelling and treatment. The network is composed of academics, researchers and experts from the waste management, wastewater treatment and agricultural industries.

The report has been funded through the Shanks First fund by EB Milton Keynes Ltd (an Environmental Body registered under the Landfill Tax Credit Scheme), in partnership with Anglian Water Services Ltd, Northumbrian Water Ltd, Severn Trent Water Ltd and Thames Water Plc. The authors would like to express their sincere thanks to these funders, and also to those who offered guidance and advice on the report contents.

List of Task Group Members

Prof. Christophe Guy
ECOLE Polytechnique, Canada

Prof. Franz-Bernd Frechen
University of Kassel, Germany

Prof. Lawrence Koe
Nanyang Technological University, Singapore

Dr Richard Stuetz – Task Group Co-ordinator
Cranfield University, UK

Prof. Herman Van Langenhove
University of Ghent, Belgium

Jay Witherspoon
CH2M Hill, USA

Acknowledgements

Various people have contributed to this technical report both scientifically and financially. The authors wish to acknowledgment the contributions from the IWA Task Group, the support from EB Milton Keynes Ltd and the contributions by Anglian Water, Severn Trent Water, Thames Water, United Utilities and Yorkshire Water which made this Entrust project possible.

Additional thanks are due to the working group of Matt Lovell, Peter Winter, Michael Cotter, Kathryn Darke, Son Lee, Richard Barnes and Richard Radley for their useful discussions during the preparation of the technical report.

1

Introduction

Effective odour management requires measurement. Information on the sources of odours, their composition, their magnitude, and their impact on the surrounding community is required in order to assess control and minimisation options. Unfortunately, odours are difficult to measure. Human response to odours is highly subjective – different people find different odours offensive, and at different concentrations. Sampling and measurement are further complicated by the fact that most odorous emissions consist of many individual odorous components, and the overall odour of complex mixtures cannot easily be predicted. For these reasons, there is no universally accepted method for the quantification of odours, and odour measurement has often been regarded as an art as opposed to a science.

Many odour measurements cannot be carried out in the field and the odour sample must be transported to a laboratory. For the measurement to be meaningful, it needs to be representative of the odour as experienced in the field. Hence the sample must be preserved in some way so as not to degrade or change its character in the time elapsed between collection and measurement.

Where odour sources are under investigation, the rate at which odours are emitted from the source is important. Measurement of emission rates is often difficult and it requires special techniques, but it is essential if the impact of the source is to be accurately assessed.

Hence there are three areas to consider (Figure 1.1). First, the sample must be collected from the source in a manner which allows the emission rate to be derived. Second, the sample must be preserved during transit from source to laboratory. Third, there is the odour measurement itself.

There are a variety of emission source types encountered in waste management industries. Solid and liquid surfaces present different challenges, as do static and dynamic surfaces. For ease of measurement, emission surfaces would be small, relatively homogeneous, solid and static. Unfortunately, these are relatively rare. Liquid surfaces may be relatively homogeneous but can be dynamic due to mixers, aerators, or the action of the wind. The more difficult sites would have very dynamic surfaces, cover large areas, and have considerable spatial and temporal variations in emission rates. Active landfill sites fit within this category.

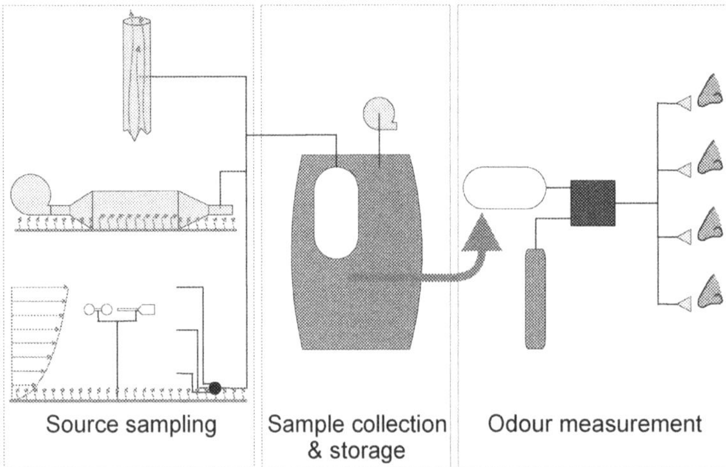

Figure 1.1. Areas to consider in odour sampling and measurement.

This report is a review of source sampling, sample collection and storage, and odour measurement. It has been written principally with waste management applications in mind, with an assumption that odours are being sampled for nuisance prediction or abatement purposes. A large proportion of the references relate to wastewater or agricultural odours, due to a relative lack of published work regarding odour sampling from waste management sites. There are, however, several examples of sampling of landfill gas emissions, which are directly relevant to odour sampling. Many of the techniques reviewed are generally applicable to waste management, as well as more widely to fields such as agriculture, wastewater treatment and industry.

The report has highlighted that there are many different techniques available for source sampling, sample collection and preservation, and odour measurement. Of these, there is no universally 'best' technique that will be suitable for the variety of emission situations likely to arise in the waste management industry. Many of the techniques may appear deceptively simple, but have the potential to interfere with emission mechanisms and give misleading results. The report aims to highlight the sources of variability in the different techniques to make the practitioner aware of their implications, indicate established practice and also to highlight further research needs. To this end, the different techniques are discussed in detail, and summary tables are provided which highlight the advantages, disadvantages and potential applications of each technique.

2

Odour measurement

2.1 INTRODUCTON

In considering odour measurement, it is important to distinguish between odorants and odours. Odorants are the compounds responsible for imparting an odour, whereas an odour is the perceived effect of the odorants as detected and interpreted by the olfactory system. By analogy, for noise measurement, the sound pressure level is responsible for imparting the sensation of sound, and the sound is the perceived effect as interpreted by the auditory system. Noise measurement is relatively easy, as there is a clear linkage between sound pressure levels and sound intensity as perceived by humans. A noise assessment would utilise the analytical measurement of sound pressure levels rather than the sensory measurement of audiometry. Unfortunately, for odours the linkage between odorant properties and odour perception is not clear, due to the lack of a comprehensive theory of olfaction. Two broad classes of odour measurement are widely used as a result: analytical measurements, referring to odorants, and sensory measurements, employing human subjects, relating to odours.

2.2 SENSORY MEASUREMENTS

Sensory measurements employ the human nose as the odour detector. In this sense, they relate directly to the properties of odours as experienced by humans. The problems of complex mixtures, interactions between components and detectability below the threshold of smell become irrelevant as the 'total effect' of the overall odour is measured.

There are many factors other than the properties of the odour sample itself that may influence the perception of an odour. Principal amongst these is the variability in the sense of smell between different observers. This is usually minimised by using a panel of several observers, and averaging the individual results. Great care must be taken in the presentation

© IWA Publishing. *Sampling for the Measurement of Odours*. P. Gostelow, P. Longhurst, S. A. Parsons and R. M. Stuetz. ISBN: 1 84339 033 7

of samples to observers to give repeatable results. Factors such as the order in which samples are presented, the environment in which the testing takes place, and the flow rate of the carrying gas stream are all important.

Before discussing odour measurement techniques, it is useful to consider the parameters that can be measured.

2.2.1 Odour parameters

Environmental odours are generally recognised to have four sensory parameters:

- threshold concentration – the lowest concentration at which an odour can be detected (detection threshold) or recognised (recognition threshold)
- intensity – the perceived strength of the odour
- character – what the odour smells like
- hedonic tone – the degree of pleasantness or unpleasantness associated with the odour

The concept of detection and recognition threshold concentrations arises because odours may be detectable by human subjects, but at concentrations too low to allow identification of the odour. Recognition thresholds are typically higher than detection thresholds by a factor of 1.5 to 10 (Dravnieks and Jarke, 1980).

The odour detection threshold is fundamental to the measurement of odour concentration. This is measured by diluting an odour sample with odour-free air until it is reduced to its detection threshold concentration. The odour concentration is then expressed as the number of dilutions required to achieve this.

The intensity of an odour refers to the strength of an odour as perceived by a human subject. Intensity is usually represented using a subjective category scale (i.e. faint – moderate – strong), by subjective magnitude estimates (i.e. odour A is twice as strong as odour B) or by reference to a specific odorant, whose concentration is adjusted until the reference and test odorant have the same perceived intensity.

The character of an odour is subjective and described by analogy. For example, the character of hydrogen sulphide is similar to the odour of rotten eggs, mercaptans smell similar to rotten cabbage, amines smell fishy.

The hedonic tone of an odour is the degree of pleasantness or unpleasantness associated with an odour. Typically this will be represented using a numeric scale depicting most pleasant at one end and most unpleasant at the other. Often, negative values are used to represent unpleasant odours and positive values represent pleasant odours.

There are several links between the odour parameters. In general, intensity increases logarithmically with odour concentration as characterised by Steven's law or Weber-Fechner's law (Wright, 1982). The character of an odour influences its hedonic tone. Both character and hedonic tone can be influenced by intensity.

Although there are several odour parameters which can be measured, for odour nuisance situations measurement of odour concentration is usually considered adequate. Dose-response studies have indicated that hedonic tone has a slight but significant effect on annoyance (Miedema *et al.*, 2000), but for the purposes of nuisance prediction, it is assumed that all odours are equally annoying. It has been hypothesised that the odour quality is of limited significance if the odour is seen as 'out of place' in a residential area (Miedema and Ham, 1988), which is consistent with the observation that normally pleasant odours can become unpleasant if involuntary exposure occurs on a regular basis.

A list of common odorants associated with waste treatment, their characters and threshold concentrations is given in Appendix A.

2.2.2 Sensory measurement techniques

Sensory measurements can be considered as subjective, where the nose is used without any other equipment, or objective, where the nose is used in conjunction with some form of dilution apparatus (Koe, 1989).

Parameters which may be measured subjectively include odour character, hedonic tone and intensity. Indeed, for character and hedonic tone, no objective techniques are available.

Intensity is often measured subjectively, typically using ordinal category scales. These usually employ three to ten categories, but variations on a six category scale shown below are common (Cheremisinoff, 1988; Frechen, 1994):

0 = no odour perceivable
1 = barely perceivable
2 = faintly perceivable
3 = clearly perceivable
4 = strong
5 = very strong

Interpretation of subjective sensory odour measurements is difficult due to the inherent variation of odour perception even for well trained personnel (Koe, 1989). Particular difficulties arise when comparing results from different studies, particularly where different category scales have been used. Because of their inherent variation, their use for regulatory purposes is limited.

The main advantage of subjective measurements is that they require no special equipment and can be obtained quickly at low cost. They are useful for preliminary assessments and can raise awareness of odours among those taking part in the measurements. Heroux *et al*. (2001) and van Langenhove and Broeck (2001) describe methodologies for conducting subjective odour measurements, and Finnigan (1998) gives an example of the use of subjective measurements in site operation and awareness raising at sewage treatment works.

Objective sensory measurements employ the nose in conjunction with an olfactometer, an instrument which dilutes the sample with odour-free air. There are two categories of dilution-related measurement techniques. Suprathreshold olfactometry is a measure of odour intensity in which the sample odour is compared to a reference odour and the result is expressed as an equivalent concentration of the reference gas. The more common method is threshold olfactometry which gives a measure of odour concentration by diluting the sample until it can only just be detected (the threshold concentration). In both cases, the aim of the olfactometer is to remove subjectivity from the measurement. The panellist has to make no judgement about the odour other than whether it can be detected (threshold olfactometry) or whether it is more or less intense than a reference odorant (suprathreshold olfactometry).

Dilution may be static or dynamic. Static dilution involves the mixing of fixed volumes of odorous and odour-free air, whereas dynamic dilution involves the mixing of known flows. Dynamic dilution is superior to static dilution as the effects of sample adsorption to the internal surfaces of the instrument are minimised (Dravnieks and Jarke, 1980), and static instruments or techniques have largely been superseded (Koe, 1989; Brennan, 1993). An additional advantage of dynamic olfactometers is that the sample can be delivered to the sniffing port at a constant flow, a factor which has been shown to improve repeatability of results (Duffee and Cha, 1980; Koe, 1989; Schulz and van Harreveld, 1996).

2.2.2.1 Threshold olfactometry

The basis of threshold olfactometry is the dilution of a sample with odour-free air until it is reduced to its threshold concentration. A simple flow scheme for a dynamic olfactometer is shown in Figure 2.1, and a six-station olfactometer is shown in Figure 2.2.

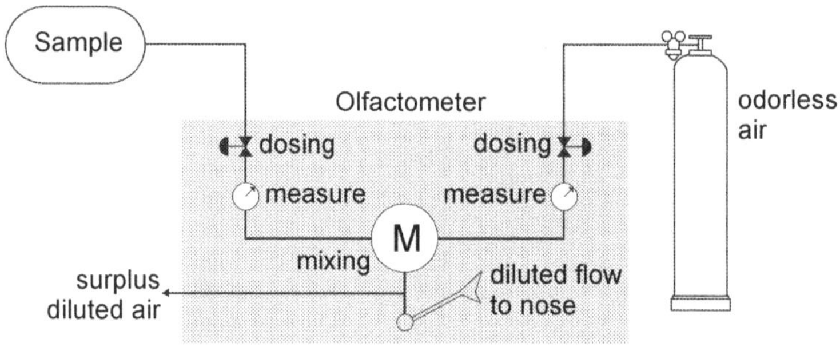

Figure 2.1. Flow scheme of dynamic olfactometer (after Frechen, 1994).

The essential elements of a dynamic olfactometer are (Koe, 1989):

- air pumps for drawing in the odorous samples and to provide the range of flow rates
- flow meters for accurate measurement of airflow rates
- mixing chambers to mix the flows of odorous and odour-free air
- deodorising chambers, usually employing activated carbon, or an alternative supply of odour-free air
- control valves for accurate control of airflow rates
- connecting tubes for conveyance of flows between various parts of the olfactometer and to the sniffing port
- sniffing port (Figure 2.3), usually a face mask, hood or funnel where the sample is delivered to the observer's nose

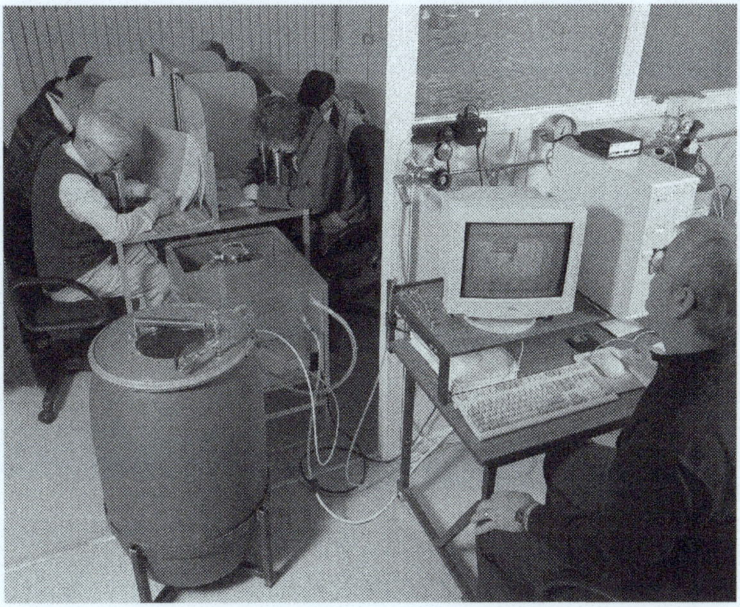

Figure 2.2. A six-station forced choice olfactometer (courtesy of Silsoe Research Institute, UK).

Figure 2.3. Sniffing ports for a forced choice olfactometry (courtesy of Silsoe Research Institute, UK).

Odour concentrations derived by threshold olfactometry are dimensionless, mathematically expressed as:

$$C = \frac{Q_0 + Q_f}{Q_0} \tag{2.1}$$

where: Q_0 = flow rate of odour sample
Q_f = flow rate of odour-free air required to odour at sample port to threshold

Concentrations derived by threshold olfactometry may be termed Threshold Odour Numbers (TON) or Dilution to Threshold (D/T) ratios. They represent the number of dilutions of the sample necessary to reach the odour perception threshold. Although these values are dimensionless, it is becominging increasingly common to envisage them as physical mass concentrations, in which case they tend to be termed odour units per cubic meter (ou m^{-3}) (Frechen, 1994; Koe, 1989). This has advantages in terms of consideration of odour mass loadings or emission rates, but it must be remembered that it is an arbitrary designation, and has been criticised by some (i.e. Walker, 2001).

There are a number of variables involved in olfactometry which will affect measurements. Key amongst these are olfactometer design, the test procedure, the differing sensitivity of observers, and the method of calculation used to determine the threshold calculation (Dravnieks and Jarke, 1980; Koe, 1989).

Olfactometer design

The materials used in olfactometer construction should not lead to sample contamination or alteration through adsorption/desorption. Low adsorbency materials such as stainless steel, teflon, tedlar or glass are used (Jiang, 1996; Stordeur *et al*., 1981), and it is also beneficial to minimise internal surface areas and provide provision for flushing or easy replacement of flow lines between samples (Dravnieks and Jarke, 1980). The effects of adsorption/desorption can also be minimised by utilising an ascending order of sample presentation, whereby the weakest samples (highest dilution) are presented before stronger samples (lower dilution).

To obtain accurate results, the flow of the odour sample and the dilution gas need to be controlled and measured accurately. Consideration should be given to the means of controlling the gas flows, typically involving the use of rotameter/fixed orifices or mass-flow controllers. These can be sensitive to back-pressure from downstream measurement devices (Jiang, 2001); careful design and calibration are required to ensure accuracy.

Because of adsorption and ingress of particulates, olfactometers require periodic calibration using tracer gases to maintain accuracy. This can involve a large number of measurements; the provision of automatic calibration capabilities is desirable (Jiang, 2001).

Test procedure

The schedule of presentation of odour samples can influence results by two mechanisms (Dravnieks and Jarke, 1980; Koe, 1989). First, if a descending order of presentation is used, olfactory fatigue or adaptation can set in, whereby a weak odour is more difficult to detect after exposure to a strong odour. Second, where dilutions occur in a strict order, panellists will begin to expect subsequent samples to be weaker or stronger and may adjust their responses accordingly.

Of the problems related to presentation schedules, olfactory fatigue and adsorption/ desorption appear to have the greatest effect. In this case it is better to utilise ascending order presentation (Dravnieks and Jarke, 1980). Ascending order presentation also has the advantage of minimising panellist exposure to potentially toxic substances in the sample.

The supply of odour-free air is important, as any residual odour in the dilution air would bias the result. Options include the use of cylinder gas or carbon-filtered room air (Dravnieks and Jarke, 1980; Stordeur *et al.*, 1981; Sweeten *et al.*, 1984). The best option is to provide a deodorised supply to the test room and to use this directly as the dilution air; this will ensure that panellists are located in an odour-free atmosphere (Cheremisinoff, 1988; Dravnieks and Jarke, 1980).

Different flow rates at the sniffing port can have a major influence on the reported odour concentration. Dravnieks and Jarke (1980) showed that where a constant flow rate was maintained, the ratio between the highest and lowest threshold concentration reported by an odour panel was 2.5. Where the flow rate was varied between 0.5 to 9 L min^{-1} the ratio was increased to 4. Duffee and Cha (1980) state that a 100-fold variation in the sniffing port flow rate can produce a more than 1000-fold variation in the reported concentration. Ideally, the flow rate should be matched to air requirements for normal breathing and sniffing, as sample flows lower than this can lead to increasing dilution with room air, leading to reduced measurement sensitivity (van Harreveld *et al.*, 1999). O'Brien *et al.* (1995) investigated the effect of flow rates between 1 – 20 L min^{-1} on the measured threshold concentrations of p-Xylene, hydrogen sulphide and n-butanol. Two different sized sniffing ports were utilised, allowing the effect of face velocity and flow rate to be separated to a certain extent. Results (Figure 2.4) show a decrease in threshold concentration (and hence increase in sensitivity) with increasing flow rate. These results suggest flow rates above 5 L min^{-1} are preferable, and O'Brien *et al.* recommended flow rates between 5 – 10 L min^{-1}.

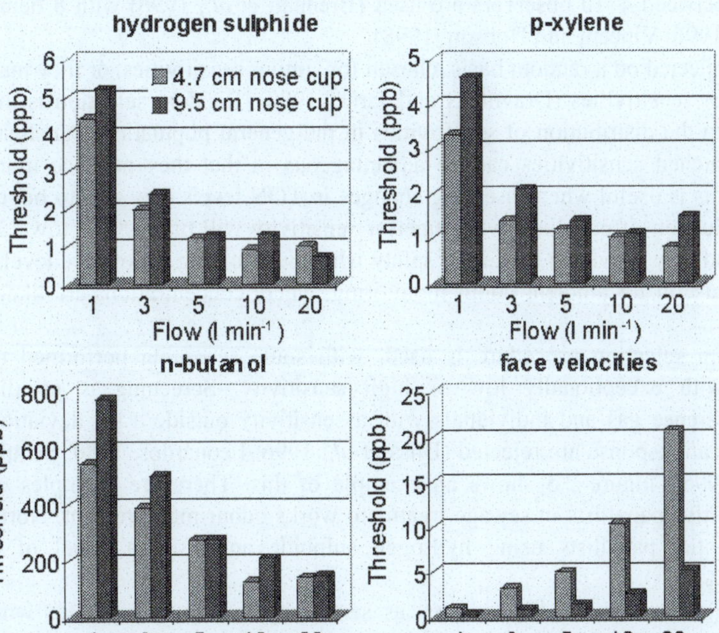

Figure 2.4. Threshold concentrations at different olfactometer flow rates (after O'Brien *et al.*, 1995).

There are two methods in common use for observers to indicate whether an odour can be detected at a particular dilution – yes/no or forced choice. In the yes/no method, a single sniffing port is used and the observer is asked: 'can you detect an odour?' In forced choice, two or more sniffing ports are used. The odour sample is presented at one port, and odour-free dilution air at the other port(s). In this case, the observer is asked: 'which port contains the odorous sample?' Forced choice olfactometry is preferred over yes/no olfactometry, as there may be a bias for some observers to say yes or no regardless of whether they can detect an odour (Brennan, 1993; Dravnieks and Jarke, 1980). An additional parameter may be applied to both forms of olfactometry, whereby the panellists are asked to indicate the degree of certainty they have in their response. This varies from guessing to certain, and gives an additional confidence measure when interpreting panellist responses.

Other elements of the test procedure can influence results. Various cues can affect a panellist's response – for example, the response of other panellists, position of valves or gauge readings on the olfactometer, differences in appearance of sniffing ports where more than one is used, and responses of the operator (Dravnieks and Jarke, 1980). It is therefore beneficial if panellists are isolated from each other, from the olfactometer controls and from the operator.

Panellist sensitivity

Sensitivity to odours is variable between different individuals, meaning that different observers will report different odour concentrations for the same sample. This effect is minimised by the use of a panel of several observers and recording the average response. The number of observers is a trade-off between resolution of results and practicality and cost. If too few panellists are used, statistical tests may lack resolution. However, if a large panel is used both the cost of measurement and the time taken to determine sample concentrations increases. Typically, between 4–10 observers are used (Brennan *et al.*, 1996) with 8 being common (Bliss *et al.*, 1996; Vincent and Hobson, 1998).

Observers may be selected on a random basis, chosen for similar sensitivities, or they may be selected for superior sensitivities (Dravnieks and Jarke, 1980). Random selection is the best basis for reflecting the distribution of sensitivities in the general population. Panellists chosen for closely matched sensitivities can be advantageous in that they produce more reproducible results; this is useful when assessing changes in TON levels due, say, to odour control equipment. Selection of panellists with superior sensitivity will bias results towards higher TON values, and may provide a margin of safety when determining annoyance levels. This approach is also useful for ambient odour monitoring, where concentrations are likely to be low.

Generally, a random selection procedure is used, with some screening performed to remove individuals with exceptionally low or high sensitivity. Screening is usually performed using a reference gas and individuals with a sensitivity outside a set deviation from the panellists' mean response are rejected (Bliss *et al.*, 1996; Leonardos, 1980; Schulz and van Harreveld, 1996). Figure 2.5 shows an example of this. There are examples of selection of more sensitive panellists in sewage treatment works odour measurement. North (1979) screened potential panellists using hydrogen sulphide and used a panel of 6 comprising the top 80%.

Prior experience of observers can bias results, as sensitivity to odours increases with experience and training (Cain, 1980; Leonardos, 1980). Panellists who have previous experience of olfactometry may bias results towards higher TON values. Bias may also occur as a test progresses, although the use of screening with a reference gas will reduce the effect, as it familiarises panellists with the test procedure before the odour samples are tested.

Calculation of TON

When the TON values for individual panellists have been determined, it is necessary to derive some average TON value to represent the panel as a whole. This is commonly derived using either the median or the geometric mean. Tests using large numbers of individuals have suggested a log-normal sensitivity distribution. For panels of practical size the log-normal distribution is often absent and the geometric mean and the median may not coincide (Dravnieks and Jarke, 1980). Where panellists are asked to apply a level of certainty to their measurements, the interpretation of these responses can also influence results. Bliss *et al.* (1996) compared 'certain and correct' against 'guess and correct' responses in a forced-choice olfactometry experiment, and found the threshold concentration for n-butanol was derived as 19.22 ppb when derived on a guess and correct basis, against 54.59 ppb for certain and correct.

2.2.2.2 Standards for olfactometry

The number of variables which can influence the results obtained by threshold olfactometry have led to poor reproducibility between different laboratories with ranges of 1 to 3 orders of magnitude difference being common (Schulz and van Harreveld, 1996). This has led to olfactometry having a poor reputation and makes comparisons between different odour measurements difficult. Recognition of these problems has led to the development of standards for olfactometry.

Development of olfactometry standards has been influenced largely by inter-laboratory studies carried out in the Netherlands (Harssema and Heeres, 1992; Schulz and van Harreveld, 1996; van Harreveld *et al.,* 1999). Several rounds of comparisons were carried out, with the following developments being made (Shulz and van Harreveld, 1996):

- standardised olfactometer calibration using tracer gases
- standardisation of odour flow at sniffing port to 20 L min^{-1}
- standardisation of calculation method
- minimum panel size of 8 members
- strict screening of panel members for sensitivity to a standard odorant (n-butanol)
- introduction of a common anchor-point for the odour unit, with reference value of 1 ou m^{-3} = 20 ppb/v n-butanol

The major increase in repeatability occurred in the last round where panel screening was introduced. For a panellist to be accepted, they must, on the basis of at least 12 trials, estimate the threshold of a 40 ppb/v n-butanol standard to be between 20–80 ppb and must have a standard deviation of the logarithm of the estimates less than 0.916. The inter-laboratory studies led to the introduction of the Dutch NVN2820 standard.

The draft European CEN (CEN, 1999) standard has largely been developed from the Dutch standard, but differs slightly (Schulz and van Harreveld, 1996):

- improved quality of panellists' responses by addition of 'guess' / 'inkling' / 'certain' criteria to forced-choice decision
- allowance of yes/no methods as well as forced-choice
- shifting of reference odour unit to 1 ou m^{-3} = 40 ppb/v n-butanol

The key points of the CEN standard are (van Harreveld *et al.,* 1999):

- *equipment calibration*: tracer gas calibration must show dilution fraction to be within 20% of the expected value (95 percentile basis)
- *panellist selection*: at least 12 individual threshold estimates of a 40 ppb/v n-butanol standard must show a geometric mean of between 20–80 ppb/v, and the antilog of the standard deviation based on the logarithms of the individual estimates to be less than 2.3
- *panel size*: A minimum panel size of 4 is allowed after retrospective screening, although larger panel sizes are recommended
- *accuracy*: the difference between the geometric mean of the panellist estimates of a 40 ppb/v n-butanol standard and 40 ppb/v should be within a factor of 1.6 (i.e. group estimates of the 40 ppb/v standard should lie within 25–64 ppb)
- *repeatability*: the difference between two subsequent concentration estimates of a single sample should be less than a factor of 3

A glossary of terms and definitions used in the CEN standard is included in Appendix B.

Measurements of these criteria at an olfactometry laboratory (Silsoe Research Institute, Bedfordshire, England) are shown in Figures 2.5, 2.6 and 2.7. As can be seen, the portion of the population from which panellists are selected represents those with a sensitivity close to, or more sensitive than, the average.

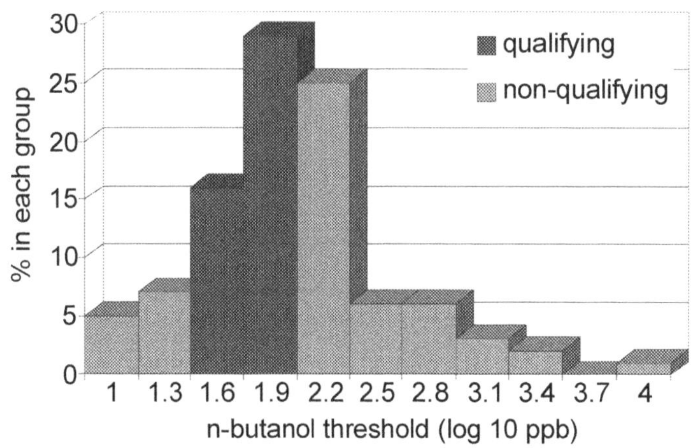

Figure 2.5. Distribution of individual threshold estimations for n-butanol at SRI (after Sneath, 2001).

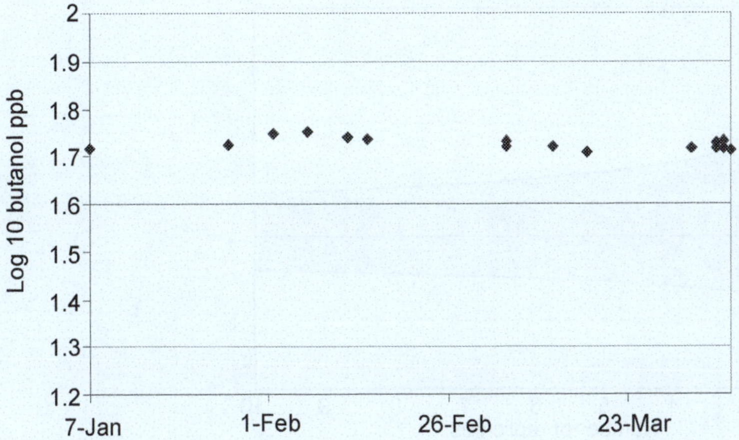

Figure 2.6. Three-month history of average panel threshold at SRI (after Sneath, 2001).

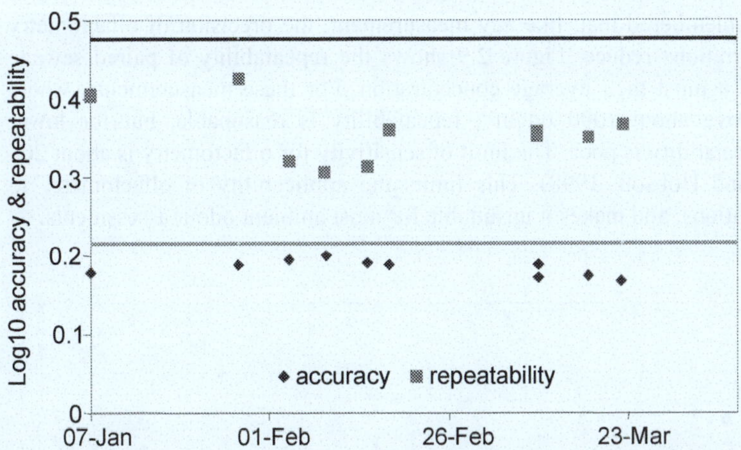

Figure 2.7. Three-month history repeatability and accuracy at SRI (after Sneath, 2001).

The standardisation of olfactometry has greatly improved the repeatability of odour measurement. It must be remembered that, as a sensory measurement, it can never offer the levels of repeatability and accuracy characteristic of analytical measurements. The CEN standard allows a repeatability factor of 3. This has implications for the confidence in olfactometry results and the number of samples required. Figure 2.8 shows the 95% confidence limits for an odour sample of 1000 ou_E m^{-3} resulting from an increasing number of samples. As can be seen, there is a significant reduction in the confidence limits for sample sizes above 2, and little benefit in increasing sample numbers beyond 4. As a result, triplicate or quadruple samples are recommended as a balance between cost and precision (Sneath and Clarkson, 2000).

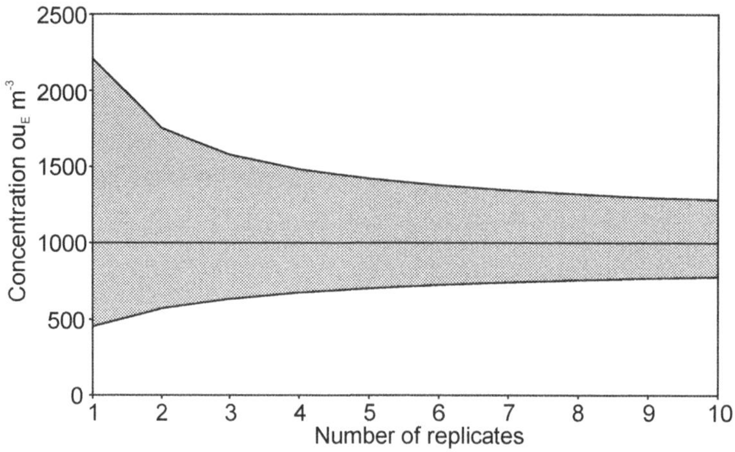

Figure 2.8. 95% confidence limits with replicate samples (after Sneath and Clarkson, 2000).

It should also be remembered that, like any measurement, the precision of olfactometry will reduce as concentrations reduce. Figure 2.9 shows the repeatability of paired sewage odour samples plotted against their average concentration. For these measurements, where concentrations are above about 1000 ou m^{-3}, repeatability is reasonable, but for lower concentrations the repeatability is poor. The limit of sensitivity for olfactometry is about 20–50 ou m^{-3} (Vincent and Hobson, 1998). This limits the applicability of olfactometry to relatively odorous situations, and makes it unsuitable for most ambient odour assessments.

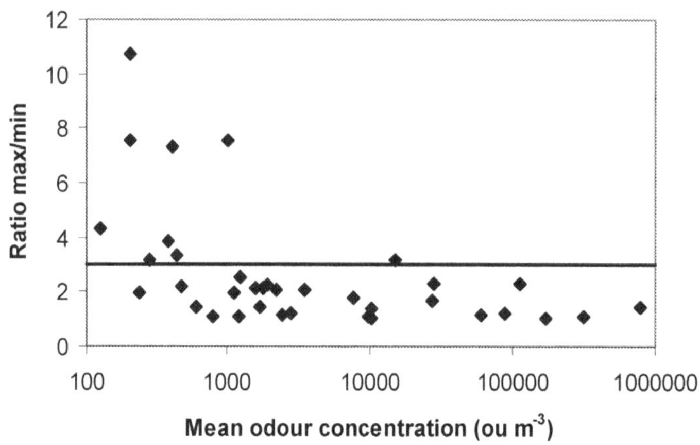

Figure 2.9. Repeatability of paired sewage odour samples (after Gostelow and Parsons, 2000).

2.3 ANALYTICAL MEASUREMENTS

Analytical measurements concern the physical or chemical properties of the odorous compounds, although the most common measurement made by far is odorant concentration. Analytical measurements have the advantage of objectivity, repeatability and accuracy. They can also be related directly to theoretical models describing chemical transformations or mass transfer. They do, however, have a number of disadvantages (Brennan, 1993; Preti *et al*., 1993; Stordeur *et al*., 1981; Young, 1984):

- Most environmental odours are complex mixtures of dozens of components. This complicates analysis considerably, which usually requires a separation technique prior to analysis.
- Odorants may be present in very small concentrations. The limit of analytical detection may be below the threshold of smell. Non-odorous compounds will also be present in the sample in much larger concentrations than the odorants.
- It is difficult to relate analytical measurements to the intensity of odour as perceived by a human observer. This is especially the case for mixtures, as interactions between different odorants may lead to synergistic or antagonistic effects.

Because most environmental odours are mixtures, separation techniques such as gas chromatography are often used for qualitative or quantitative measurements of odour composition. Gas chromatography relies on the differential absorption of the sample components on the wall of a long chromatography column, such that the different molecules emerge from the end of the column at different times. Figure 2.10 shows a schematic of a gas chromatography system. In some cases, it is useful to complement gas chromatography with sensory techniques, whereby some of the chromatography eluant is diverted to a sniffing port, allowing continuous olfactory sampling to be carried out. Olfactory-gas chromatography allows focusing of the analytical effort to the parts of the eluent which are odorous or the components having characteristics of the overall odour of the sample (North, 1979; Preti *et al*., 1993).

Gas chromatography is a complex technique, and great care is required, particularly in the areas of sample collection, choice of chromatography column, and choice of detector. A detailed description of gas chromatography is beyond the scope of this report. Hobbs (2001) provides an excellent review of odour analysis by gas chromatography.

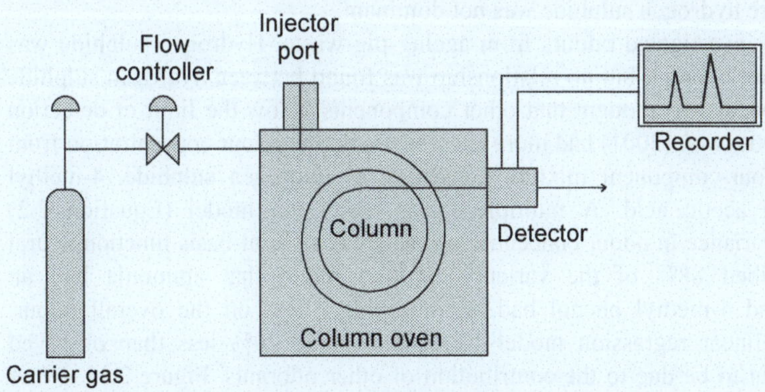

Figure 2.10. Schematic of a gas chromatography system.

It is often found that, although many odorants are present for a particular odour source, a single odorant is dominant. This can be very useful, as a measure of the single odorant can be indicative of the odour as a whole. It is particularly advantageous if the dominant odorant is easy to measure. Hydrogen sulphide is often used as a 'marker' odorant for these reasons. Hydrogen sulphide is a dominant odorant for many waste processes, and is particularly associated with sewage treatment and animal wastes. It can be problematic at landfill sites, especially if these have received gypsum wastes (Anon, 1998). It can be measured down to low parts per billion concentrations using hand-held instrumentation, allowing many measurements to be made in a short period of time without the need for sample preservation and transportation to an off-site facility. Gostelow and Parsons (2000, 2001) describe hydrogen sulphide measurements applied to sewage treatment and Hobson and Yang (2001) describe hydrogen sulphide mapping, a data visualisation technique employing field hydrogen sulphide measurements.

A key issue in analytical measurements is the relationship between the concentration of odorants, and their effect on odours as perceived by humans. For single odorants, odorant and odour concentration would be expected to be directly related since odour concentration measurements are expressed as multiples of the odorant threshold concentration. Hence, if an odorant has a threshold concentration of 10 ppb, 1 odour unit equals 10 ppb, 10 odour units would equal 100 ppb and so on. For mixtures of odorants, the relationship is less clear as synergistic or antagonistic effects can occur. Recent studies on odour mixtures have suggested a range of effects from simple additivity (Patterson *et al.*, 1993), synergistic additivity (Laska and Hudson, 1991), and partial additivity or suppression (Laing *et al.*, 1994; Hobbs *et al.*, 2001).

Where one or a few odorants are identified as dominant for certain odour sources, relationships can be formed between odorant and odour concentration. Several authors (Gostelow and Parsons, 2000; Hobbs *et al.*, 1999, 2001; Noble *et al.*, 2001) have performed odorant/odour correlations for a range of wastes.

Gostelow and Parsons (2000) performed correlations between hydrogen sulphide and odour from sewage odours. Odour concentration was related to hydrogen sulphide concentration using a power law relationship. Reasonable correlations were found for sludge treatment processes, where it is expected that hydrogen sulphide would be the most dominant odorant. A less strong relationship was found for preliminary treatment, where the contribution of hydrogen sulphide to the overall odour was less. No correlation was found for aerobic treatment, where hydrogen sulphide was not dominant.

Hobbs *et al.* (1999) investigated odours from ageing pig waste. Hydrogen sulphide was found to be the dominant odorant, but no relationship was found between hydrogen sulphide and odour concentration. It was thought that other components below the limit of detection influenced results. Hobbs *et al.* (2001) had more success predicting odour concentration from pig manure using a four-component mixture consisting of hydrogen sulphide, 4-methyl phenol, ammonia and acetic acid. A multiple linear regression model (Equation 2.2) described 74% of the variance in odour concentration, whereas a radial-basis function neural network model described 88% of the variance. It was found that ammonia had an insignificant effect, and 4-methyl phenol had an inhibiting effect on the overall odour. Predictions using the linear regression model were on average 20% less than observed values; this was thought to be due to the contribution of other odorants. Figure 2.11 shows the comparison between observed and predicted values for the linear regression model.

$$C_{ou} = 950 + 2630C_{H2S} - 25617C_{FMP} + 179C_{acetic} \tag{2.2}$$

where: C_{ou} = odour concentration, ou m^{-3}
C_{H2S} = hydrogen sulphide concentration, mg L^{-1}
C_{FMP} = 4-methyl phenol concentration, mg L^{-1}
C_{acetic} = acetic acid concentration, mg L^{-1}

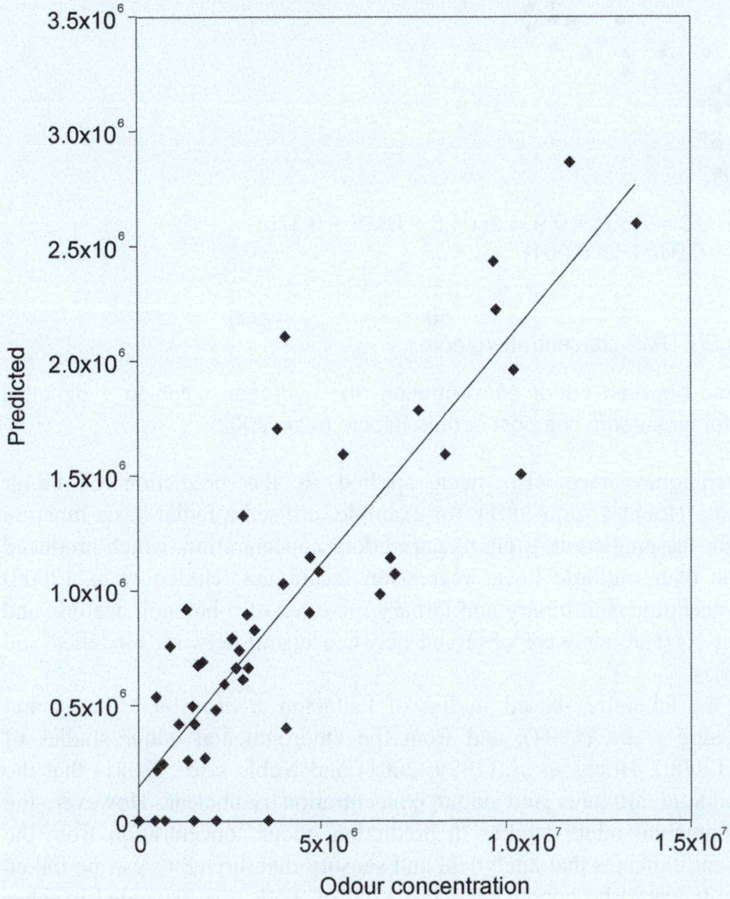

Figure 2.11. Comparison between predicted and observed values for pig manure odours using regression model (Equation 2.2) (Hobbs *et al.*, 2001).

Noble *et al.* (2001) investigated odours from mushroom composting emissions. The mushroom compost odours were taken from 11 sites, all using wheat straw, poultry manure and gypsum, plus horse manure on some sites, in different proportions. It was found that odour concentration could be predicted using hydrogen sulphide and dimethyl sulphide, using Equation 2.3. The resulting correlation is shown in Figure 2.12.

$$\ln C_{ou} = 7.601 + 0.934 \ln(C_{H2S} + C_{DMS} + 0.375) \tag{2.3}$$

where: C_{ou} = odour concentration, ou m^{-3}
C_{H2S} = hydrogen sulphide concentration, mL L^{-1}
C_{DMS} = dimethyl sulphide concentration, mL L^{-1}

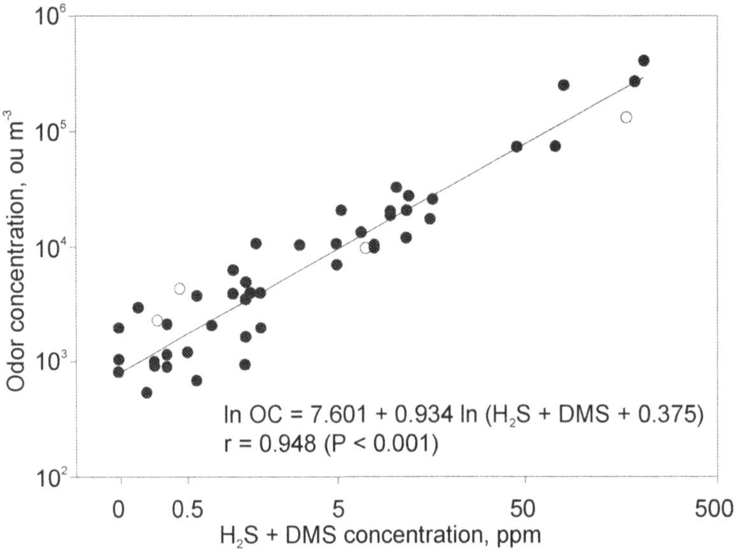

Figure 2.12. Relationship between odour concentration and hydrogen sulphide + dimethyl sulphide concentration for mushroom compost odours (Noble *et al.*, 2002).

Neural network techniques have also been applied to the prediction of odour concentration or intensity. Hobbs *et al.* (2001), for example, utilised a radial-basis function neural network model for the prediction of pig manure odour concentration, which produced an improved correlation over multiple linear regression techniques. Hudon *et al.* (2000) applied neural network techniques to binary and ternary mixtures of n-butanol, acetone and ethyl mercaptan. Strong correlations were observed between neural-network modelled and measured odour intensities.

It is apparent from the laboratory-based studies of Patterson *et al.* (1993), Laska and Hudson (1991), and Laing *et al.* (1994), and from the environmental odour studies of Gostelow and Parsons (2000), Hobbs *et al.* (1999, 2001) and Noble *et al.* (2001) that the relationship between odorant mixtures and odour concentration is unclear. However, the successes of the environmental odour studies in predicting odour concentration from the dominant odorants present indicates that analytical and sensory measurements can be linked for certain odour types. It would be beneficial if this kind of work was expanded to other odour types.

2.4 CHOICE OF ODOUR MEASUREMENT METHOD

As has been discussed, there is no universally applicable odour measurement and a choice between the different methods needs to be made. Tables 2.1 and 2.2 set out the advantages, disadvantages and potential applications of the different sensory and analytical techniques. In many cases, it will be advantageous to combine two or more different techniques, such as GC-MS and olfactometry. It is likely that this approach will become more common as improved linkages between odorants and odours are developed.

Table 2.1. Advantages, disadvantages, and potential applications of sensory measurement techniques.

Technique	Advantages	Disadvantages	Applications
All	Measurements relate to odours as perceived by humans.	No quantitative information on odour composition. Can be difficult to relate to liquid-phase odorant concentrations or emission mechanisms. Lack of sensitivity and accuracy when compared to many analytical methods.	Annoyance prediction; Situations where odours are complex or contain unknown odorants.
Subjective intensity measurement	Require no special equipment. Rapid. Low cost. Can be carried out on-site.	Very large margin of error in results. High risk of olfactory fatigue. Difficult to compare results of different studies.	Preliminary assessments. Awareness raising. Community surveys.
Threshold olfactometry	Becoming standardised. Relatively accurate. Precision and repeatability limits known.	Requires carefully designed and operated dilution apparatus (olfactometer). Requires several people to form odour panel. Care required over sample presentation and results calculation. Results are slow to obtain. Expensive. Cannot easily be performed on-site – requires sample collection and storage. Relatively small number of measurements possible per day.	Objective source measurements. Ranking of odour sources. Assessment of odour control efficiency. Dose-response studies.

Table 2.2. Advantages, disadvantages, and potential applications of analytical measurement techniques.

Technique	Advantages	Disadvantages	Applications
All	Measurements can be linked to physical/chemical models of odorant transformations/mass transfer. Potentially higher accuracy and precision than sensory measurements.	Results are difficult to relate to odours as perceived by humans. Many odorants may be present at very low concentrations.	Analysis of odorant transformations and emission. Design of odour control equipment.
Gas chromatography	Qualitative/quantitative analysis of complex odour samples. Identification of unknown odorants with suitable detectors. Can be used in conjunction with sensory measurements (olfactory GC).	Complex. Expensive. Difficult to perform on site – sample collection and storage required. Targeted pre-concentration of samples often required.	Identification of unknown odorants. Quantitative analysis of complex mixtures. Development of analytical-sensory links.
Odour marker measurements (e.g. hydrogen sulphide)	Can have good links with sensory measurements. Many markers can be rapidly measured in the field using hand-held equipment. Can be easy and quick to measure.	Limited to relatively simple odours or those where relatively few components describe most of the sensory effect.	Odour surveys. Micrometeorological measurements. Studies of spatial and temporal odour variations. Receptor measurements.

3
Source sampling

3.1 INTRODUCTION

The method used for odour sampling is as important as the method used for measurement. There are two main issues in odour sampling. The first issue is the collection of the odours from the source in a controlled manner, and the second is the transfer of the sampled odours to the measuring instrument. In many cases, the measuring instrument will be remote from the sampling site, so sample storage becomes an issue. This chapter concerns source sampling. Sample collection and storage are detailed in Chapter 4.

When sampling odour sources, it is usually insufficient to measure odour concentration in isolation. It is also necessary to measure the airflow associated with the source, because in most cases the odour concentration and airflow are linked. The important parameter is the odour emission rate, this being the product of airflow and odour concentration. Emission rates are used directly as design inputs for odour control equipment, or as inputs for impact assessment by dispersion modelling.

The ease of source sampling depends on the generic type of the source. Point sources are where the odours are emitted from a single point, usually a stack. These are the easiest types of source to measure, as the emission is confined within a duct, and representative concentrations and flow rates can easily be determined. Volume sources are typically buildings where odorous air escapes either intentionally through ducts or unintentionally through doors, windows or other openings. Emission measurement is more complicated here as the emission points are less well defined and controlled. Area sources consist of surface emissions from a relatively large area. Examples are landfill sites, emissions from wastewater tank/lagoon surfaces or from slurries applied to farmland. Emission rate estimation is difficult for these sources as it is not easy to measure a representative odour concentration, and there is no well defined airflow.

3.2 POINT SOURCES

Point sources are usually the controlled points at which ducted odours escape to atmosphere, usually through a stack. Emission rate measurement is relatively easy for these sources, although care is needed to ensure that the average flow rate is measured accurately because velocity profiles within ducts are not uniform. Odour samples are obtained by inserting a sampling probe into the stack and withdrawing the sample from the airflow. Pre-dilution of moist, warm samples may be required to prevent formation of condensation in sample bags. This is discussed in more detail below. Flow rate is usually measured using anemometers or pitot tubes at several positions across the area of the stack, the flow rate being the product of velocity and cross-sectional area. Measurement of flow rate in ducts is a mature technique and is covered in ISO 10780 (ISO, 1994).

3.3 VOLUME SOURCES

Volume sources refer to emissions which escape or are ventilated from buildings. As for stack emissions, both airflow and odour concentration must be known. Unlike stacks, there are likely to be multiple inlets and outlets to the building which are relatively uncontrolled.

The principal problem in determining building emission rates is the determination of the ventilation rate. A full review of methods of estimating ventilation rates for buildings is beyond the scope of this report; for more information the reader is directed to works such as Etheridge and Sandberg (1996).

Ventilation rates can be measured either by the use of anemometers at inlets and outlets to the building, or by the use of tracer gases such as CO, CO2 or SF6. Tracer techniques can be used for both force and naturally ventilated buildings, whereas anemometer techniques are more suited to force-ventilated buildings (Phillips *et al.,* 2000). Care must be taken with both methods to identify and differentiate between inlets and outlets, especially for natural ventilation or forced ventilation where ventilation rates are low (Demmers *et al.,* 1999).

3.4 AREA SOURCES

Area sources present particular difficulties in sampling because, unlike point and volume sources, there is usually no well-defined airflow associated with them. Emissions from area sources are usually governed by diffusion processes, whereby a concentration gradient provides the driving force for the transfer of odorants from the soil or liquid to the air. In some cases, there may also be a significant convective element, whereby a pressure gradient provides an additional driving force. This often occurs in landfill sites, where gas is generated by the biochemical degradation of the filled waste. In some very limited cases, such as diffused-air aeration tanks, a known flow of gas may be forced through a medium, which acts to strip out odorants.

Because of these difficulties, special methods have to be employed for emission rate measurement. Two types of measurement are commonly employed:

(1) micrometeorological methods, whereby emission rate is inferred from downwind measurements of odour concentration and wind velocity;
(2) 'hood' methods, whereby a hood or wind tunnel type apparatus is placed on the emission surface and air is blown through it. The emission rate is then given by the airflow through the hood and the odour concentration of the exit air.

3.4.1 Micrometeorological methods

Micrometeorological methods give an indirect measurement of emission rate by sampling wind velocity and odour concentration downwind of the emission source (see Figure 3.1). Sampling may occur at various heights across the entire plume section, or may employ models of spatial concentration within the plume to minimise the number of measurements required. These methods are generally more suited to analytical rather than sensory odour measurements, because dispersion will often result in low odour concentrations downwind of the source, and sensory measurements may lack the sensitivity necessary to differentiate plume concentrations from background odours. In addition, some methods require a large number of concentration samples, which adds to cost and causes practical problems if olfactometry is used.

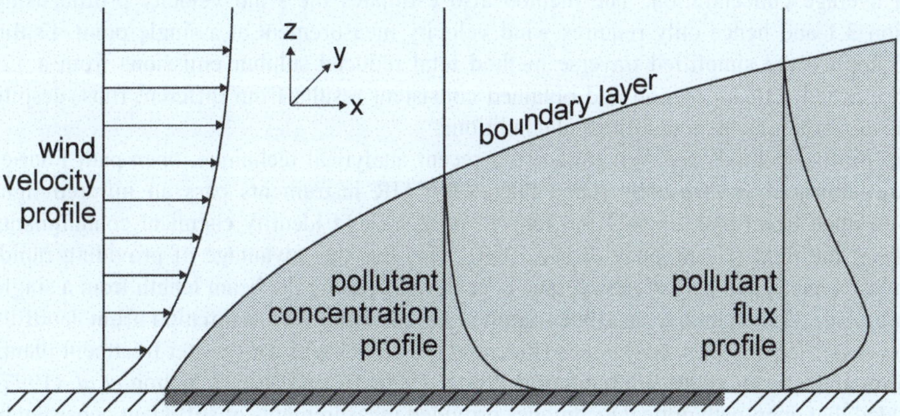

Figure 3.1. Basis of micrometeorological methods (after Esplin, 1988).

Flux profile methods

The basis of flux profile methods is the measurement of wind velocity and pollutant concentrations across the section of the plume downwind of the emission source. The product of wind velocity and concentration at a particular point gives the flux at that point. By averaging the point fluxes across the entire plume cross-section, the emission rate can be determined.

The obvious disadvantage of this method is that a large number of measurements are required at various heights to determine the spatially averaged flux. Various means have been employed to reduce the measurement requirement.

The requirement to measure wind velocity at multiple points across the plume can be reduced if it is assumed that there is no variation across the plume (the y-direction), and that the only variation is vertical (z-direction). This would require the wind velocity profile to be measured vertically at a single point only. In addition, the wind velocity profile can be approximated by a power-law relation (Irwin, 1979, cited in Esplin, 1988):

$$u_z = U_{ref}\left(\frac{z}{Z_{ref}}\right)^p \tag{3.1}$$

where: U_{ref} = wind velocity measured at a reference height Z_{ref} (commonly 10 m)
u_z = wind velocity at height z
p = an exponent depending on atmospheric stability and surface roughness

This would require the wind velocity to be measured at a single point only.

Esplin (1988) described a method which utilises continuous measurements of concentration, made at three or more elevations, whilst traversing the plume cross-section. This greatly reduces the number of point measurements required while still estimating the spatial average concentration. The method also estimates the wind velocity profile using Equation 3.1 and hence only requires wind velocity measurement at a single point. Esplin (1988) applied the simplified traverse method total reduced sulphur emissions from a 12-hectare aerated effluent lagoon and obtained consistent results from different runs, despite significant variations in meteorological conditions.

Flux profile methods are well suited to a recent analytical technique, open-path Fourier transform infrared spectroscopy (OP-FTIR). OP-FTIR instruments pass an infrared light along an open beam path up to 1 km long to measure and identify chemical contaminants directly in the field (Hashmonay *et al.,* 1999). This has the advantage of providing rapid, non-invasive measurements of the average concentration along the beam length from a single measurement. The technique has been applied to emission rate estimation from landfills (Milton *et al.,* 1995, 1997), coal mines (Piccot *et al.,* 1994), and wastewater treatment plants (Simpson and Kagan, 1990; Whitcraft and Wood, 1990; Eklund, 1999). Milton *et al.* (1997) concluded that open-path optical techniques provided measurements of sufficient consistency to quantify the extent of diurnal variations in methane emissions from landfill sites. An exciting recent development is the expansion of OP-FTIR techniques by the application of computer assisted tomography (CAT) (Piper *et al.,* 1999; Todd *et al.,* 2001). This utilises a network of OP-FTIR beams (Figure 3.2), enabling two-dimensional concentration maps of relatively large and heterogeneous sources to be obtained in a short period of time (minutes). Todd *et al.* (2001) applied this technique to ammonia emissions from a large lagoon containing swine waste. Figure 3.2 shows a sample tomographic concentration map of ammonia concentrations obtained over a 15 minute period.

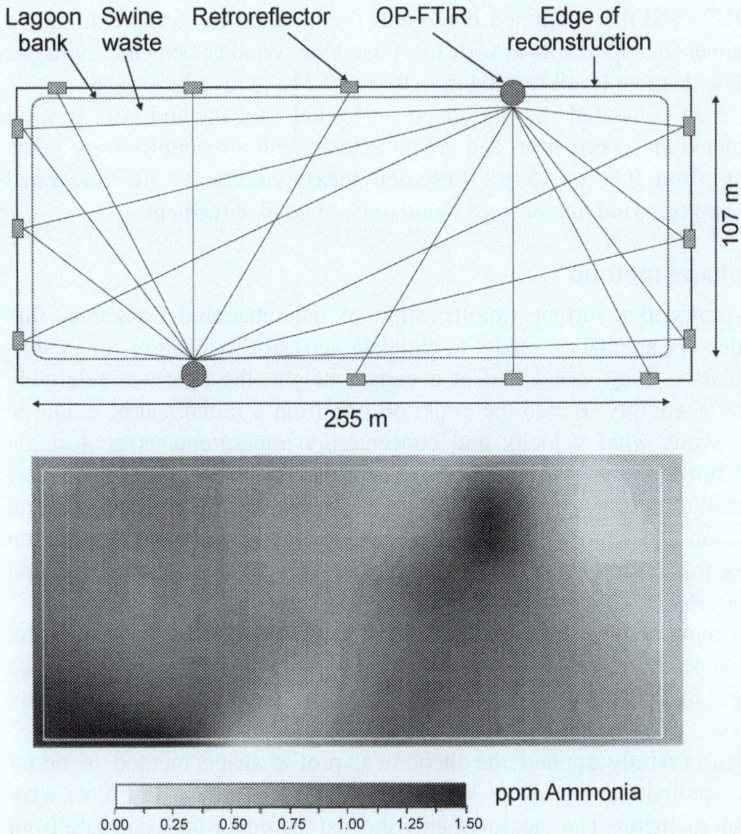

Figure 3.2. Network of OP-FTIR beams employed in computer-aided tomography techniques (top), and resulting concentration plot for ammonia concentrations from a swine waste lagoon for a 15 minute period (after Todd *et al.*, 2001).

Integrated horizontal flux method

A further simplification can be made if the emission source approximates a strip source of infinite lateral extent which is perpendicular to the wind direction, or if the emission source is radial and concentration measurements are made in the centre of the source. In this case, the system is essentially two dimensional (x and z dimensions only) and the concentration profile can be measured at a single point instead of traversing the plume. The emission rate in this case is given by (Smith and Watts, 1994a):

$$E = \frac{1}{L}\int_0^\infty C_z u_z \, dz \tag{3.2}$$

where: E = emission rate
L = fetch (downwind distance from leading edge of emission source)
C_z = pollutant concentration at height z
u_z = wind speed at height z

Denmead *et al.* (1977) used the integrated horizontal flux method to measure ammonia emissions from an experimental strip-plot 24 m wide by 500 m long. Wind velocity and ammonia concentration were measured at four levels ranging from 0.31 to 2.24 m above the ground.

Mannheim *et al.* (1995) compared the integrated horizontal flux method with a wind tunnel method for ammonia emissions from soil. Wind velocity and air samples were taken at eight levels ranging from 0.3 to 5.5 m. Emission measurements by the integrated horizontal flux method and the wind tunnel were found to be in good agreement.

Theoretical profile shape method

Wilson *et al.* (1982) provided a further simplification of the integrated horizontal flux method. Utilising a trajectory simulation model to simulate vertical flux profiles for various atmospheric stability classes, they found that at a certain height, the flux was relatively insensitive to atmospheric stability. Hence the emission rate from a radial source could be measured from simultaneous wind velocity and concentration measurements at a single height which is dependant only on surface roughness and source radius. This is a great advantage over flux profile and integrated horizontal flux methods which require several simultaneous concentration and wind speed measurements in order to derive concentration and wind speed profiles; this severely restricts their application for odour studies (Smith and Watts, 1994a; Pain *et al.*, 1991).

Wilson *et al.* (1983) compared the theoretical profile shape and integrated horizontal flux methods using ammonia volatilisation from urea applied to a 25 m radius test plot. The theoretical profile shape method was found to agree with the integrated horizontal flux method to within 8%.

Pain *et al.* (1991) successfully applied the theoretical profile shape method to odour emissions from surface applied manure. As in Wilson *et al.* (1983), circular test plots were employed, of 45 or 50 m diameter. The measurements showed the odour emission rate from applied manure decreased exponentially after spreading, and also allowed the effectiveness of different incorporation methods to be assessed.

Gaussian dispersion methods

A further micrometeorological method employs Gaussian plume dispersion theory. This was developed to predict pollutant concentrations downwind of an emission source and can therefore be used, by measuring downwind concentrations, to predict the emission rate of the source. The pollutant concentration at some point x,y,z is given by (Pasquill and Smith, 1983):

$$C_{x,y,z} = \frac{N}{\pi\sigma_z\sigma_y u}\exp\left(-\frac{y^2}{2\sigma_y^2}\right)\exp\left(-\frac{z^2}{2\sigma_z^2}\right) \tag{3.3}$$

where: σ_y = crosswind dispersion coefficient
σ_z = vertical dispersion coefficient

The dispersion coefficients are functions of downwind distance x, surface roughness and atmospheric stability. Numerous dispersion modelling software packages are available which have been developed from the Gaussian model and which can greatly simplify emission rate estimation by this method. The more advanced packages include algorithms for the influences of complex topography, building turbulence, and shorter averaging times which are potentially useful in increasing the applicability of this method to more complex sites.

The obvious advantage of this method is that the location of the measurements is relatively unimportant, provided the size and location of the emission source, and the wind direction, are known (Smith and Watts, 1994a). In addition, the restrictions applying to source geometry and wind direction for the integrated horizontal flux and theoretical profile shape methods do not apply to Gaussian methods (Smith, 1995).

Smith (1995) adapted a Gaussian dispersion model for the prediction of emission rates from areal sources. The model was found to give similar results to the theoretical profile shape method. Smith (1998) applied a Gaussian dispersion model to the prediction of hydrogen sulphide emission rates from sewage sedimentation and sludge storage tanks. Several downwind measurements were made along the plume centreline and the emission rate was determined by adjusting the model emission rate until a good fit was obtained. The closeness of fits obtained (Figure 3.3 shows an example) indicates the viability of the method.

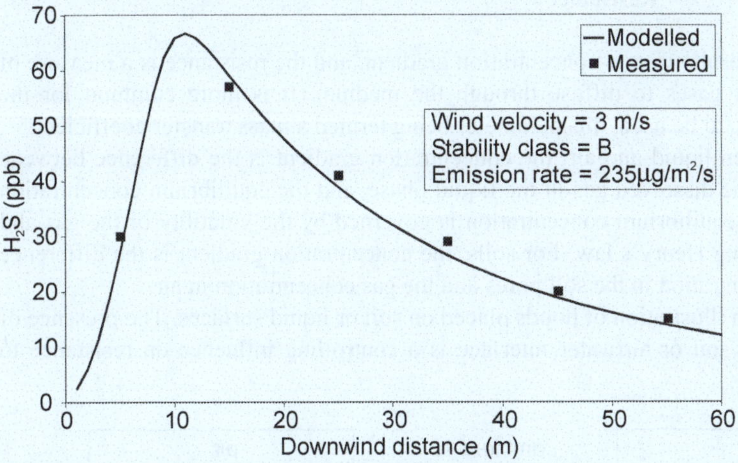

Figure 3.3. Emission rate estimate using Gaussian dispersion model (after Smith, 1998).

3.4.2 Hood methods

Hood methods give a direct measurement of emission rate. The basis of hood methods is to isolate part of the emission surface with a chamber and to force an airflow through the chamber. The emission rate is then obtained from the product of the airflow through the chamber and the odour concentration. With knowledge of the surface area enclosed by the hood, the specific emission rate can be determined (the emission rate per unit area) and can be scaled up for the entire emission surface.

Although simple in principle, hood methods require care in their implementation if they are to yield reliable results that can be related to real-world emissions. The key feature of hood methods is the enclosure of part of the emission surface, isolating it from ambient conditions and thus having the potential to alter emission rates. Care must be taken either to match hood conditions with ambient conditions, or to appreciate how hood parameters may alter emission rates. Key hood parameters which may influence emission rates include:

- flow rate through the hood
- pressure within the hood
- temperature within the hood

The flow rate through the hood is a very important controlling parameter, particularly for diffusion-driven emissions. The reasons for this can be illustrated with some reference to diffusive mass transfer theory. A detailed discussion of mass transfer theory is beyond the scope of this report, and for further information, the reader is advised to consult textbooks such as Bird *et al.*, (2002), Cussler (1997) or Thibodeaux (1996). In addition, Zhang *et al.* (2002) provide a useful analysis of transfer from soils to atmosphere from which much of the discussion below was taken.

Where mass is transferred from one location to another a driving force is required to overcome the resistance to the mass transfer. The general relationship is described by:

$$\text{Mass flux per unit area} = \frac{\text{Driving force}}{\text{Resistance}} \tag{3.4}$$

The driving force results from a concentration gradient, and the resistance is a measure of how difficult it is for gases to diffuse through the medium. It is more common for the reciprocal of resistance to be used, this parameter being termed a mass transfer coefficient.

For transfer between liquid and air, the concentration gradient is the difference between the concentration of the dissolved gas in the liquid phase, and the equilibrium concentration in the gas phase. The equilibrium concentration is governed by the volatility of the gas and can be determined using Henry's law. For soils, the concentration gradient is the difference between the gas concentration in the soil pores and the gas concentration in air.

Figure 3.4 shows an illustration of hoods placed on soil or liquid surfaces. The presence of layers close to the air/soil or air/water interface is a controlling influence on resistance to mass transfer.

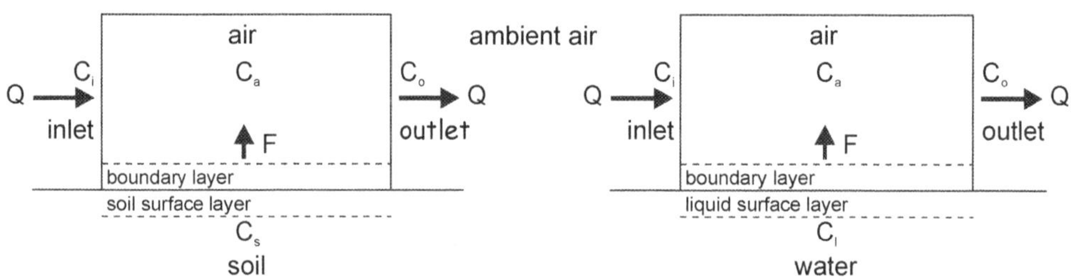

Figure 3.4. Hoods placed on soil or liquid surfaces showing presence of interface layers (after Zhang *et al.*, 2002).

Resistance to mass transfer occurs in both the boundary layer on the air-side and the surface layers in the soil or water side. The total resistance to mass transfer is the sum of these two resistances. For soils, transfer through the soil surface layer is largely governed by the porosity of the soil and the moisture content. If a soil is of low porosity, and/or has high moisture content, resistance to mass transfer in the soil surface layer is high. Similarly, for liquids, liquid-phase resistance is influenced by solubility. Highly soluble gases are mobile in the liquid phase and have low resistance to mass transfer, whereas low-solubility gases are relatively immobile and have high resistance to mass transfer.

On the air-side, transfer is controlled by the boundary layer. The governing factor here is the thickness of the boundary layer, which is controlled by surface roughness and flow velocity. For rough surfaces and high flow velocities, the boundary layer will be relatively thin and gas-phase resistance relatively low. Conversely, for smooth surfaces and low velocities, the boundary layer is relatively thick and resistance is high.

The importance of the boundary layer is largely governed by resistance in the liquid or soil phases. If resistance here is high (low porosity or saturated soils, low-solubility gases), the influence of the boundary layer is relatively unimportant. However, if resistance is low (highly porous soils, high solubility gases), the boundary layer is a controlling influence on mass transfer.

An additional influence of flow rate is apparent in liquid surfaces. As the flow rate increases, waves begin to form on the liquid surface, leading to reductions in both the liquid surface film and the air-side boundary layer. Hence the influence of flow velocity can be particularly important for liquid surfaces.

The flow rate through the hood can also influence diffusive emission rate measurements by influencing the driving force for mass transfer. If the concentration of the odorant is allowed to increase in the hood, this can serve to reduce the driving force for mass transfer and can lead to artificially low emission rates. This is graphically illustrated in Figure 3.5 which shows modelled concentration profiles under a hood placed on a cylindrical soil test cell. As can be seen, there is a significant increase in concentrations under the hood, and subsequent reduction in the concentration gradient.

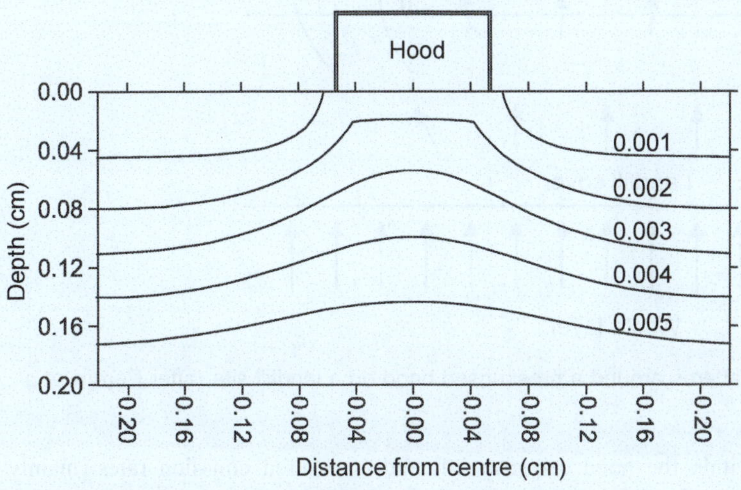

Figure 3.5. Predicted CO_2 concentration profiles for an emission hood placed on a cylindrical soil cell (after Perera *et al.*, 2002).

The flow rate through the hood can, therefore, under certain circumstances, influence the emission rate. This has been investigated by several authors. Sadek *et al.* (1998) investigated a hood method for estimating VOC emissions from liquid surfaces. They found that for liquid-phase controlled compounds (high volatility), the gas flow rate through the hood was relatively unimportant. For gas-phase controlled compounds (lower volatility), the gas flow rate was a controlling factor on the measured emission rate. Jiang and Kaye (1996) arrived at similar conclusions in a comparison of low-flow and high-flow tunnel systems. Lindberg *et al.* (2002) applied a hood method to emissions of mercury from soils. Increasing emission rate was observed with increasing flow rate. Lindberg *et al.* ascribed this to both flow-rate influence on boundary layer thickness and also reduction in the driving force due to accumulation of mercury in the gas phase at low flow velocities.

The pressure within hoods can have a great effect on emission rates where there is a significant convective component. Increased pressures can arise in a hood due to head losses caused by the flow of sweep air out of the chamber through the exit tube and associated fittings (Cooper *et al.*, 1992). If pressures are allowed to increase in the hood, the pressure gradient under the hood is altered, and convective gas flows will preferentially flow around the chamber, as shown in Figure 3.6. This can lead to artificially low measured emission rates. Alternatively, a pressure deficit can increase emissions; for example, Denmead (1979) showed that a pressure deficit of 1 mm head of water caused a 12-fold increase in the nitrous oxide emission rate from soil. Pressure effects are much less significant for emissions from water surfaces, as shown by Bliss *et al.* (1995) where an increased pressure of 8 mm head of water within a wind tunnel had little effect on ammonia emissions. Pressurisation of chambers can be reduced by using lower sweep airflow rates, large diameter ducting and fewer fittings in the exit/sampling lines.

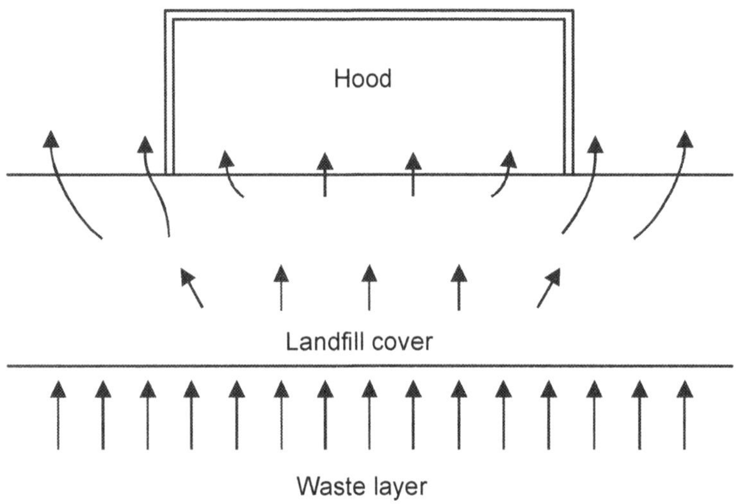

Figure 3.6. Gas flow patterns around a pressurised hood on a landfill site (after Perera *et al.*, 2002).

The temperature within the hood can also influence measured emission rates, mainly through two mechanisms. First, if temperatures become elevated in the hood, and the hood is used on moist surfaces, there is a significant risk of condensation in downstream lines or equipment, which can greatly reduce the concentration of water-soluble gases. Second, elevated temperatures within the hood may artificially increase the volatilisation of compounds sorbed to soil particles. Increased temperatures are most likely to occur in hoods made from transparent materials through a local greenhouse effect. The use of insulated chambers can minimise temperature perturbations (Cooper *et al.*, 1992).

Because the principle of hood measurements is the isolation of a portion of the emitting surface, it is essential that the hood forms a good seal with the emitting surface. This can be relatively easy for liquid surfaces, but can be difficult for coarse solid surfaces, such as many soils, vegetated surfaces, or heterogeneous surfaces such as municipal solid waste. For solid surfaces, sealing is usually accomplished by insertion of part of the chamber into the soil surface. Unfortunately, this can itself influence emission rates, as the insertion creates a channel between the hood walls and the soil surface, which acts as a path of least resistance, effectively increasing emissions at the hood edges (Cooper *et al.*, 1992). One option for reducing this problem is the use of removable skirts which are immersed into the emitting surface some days before emission measurement to minimise soil disturbance (Cooper *et al.*, 1992; Kim *et al.*, 2002). An alternative to penetration is to place the hood on the emission surface and add soil around the perimeter to effect a seal (Mosher *et al.*, 1999). Impermeable materials such as bentonite have also been employed around the base of hoods for this purpose.

There are three main types of hood methods, differentiated mainly by flow rate. These are static flux chambers, dynamic flux chambers and wind tunnels. These are discussed below.

Static flux chambers

The basis of the static flux (or equilibrium) chamber is to seal a portion of the emission surface and to allow the gas concentration to reach equilibrium within the chamber. The emission rate is then determined by the change in emission concentration over time using the equation (McGinn and Janzen, 1998):

$$N = \frac{V_C}{A_S \Delta t}\left[C_{t_2} - C_{t_1}\right] \tag{3.5}$$

where: N = Emission rate
V_C = Chamber volume
A_S = Enclosed surface area
Δt = time interval $(t_2 - t_1)$ over which concentrations C_{t1} and C_{t2} are measured

An equilibrium chamber is shown in Figure 3.7. Provision for make-up air is allowed to prevent negative pressure inside the chamber due to withdrawal of samples.

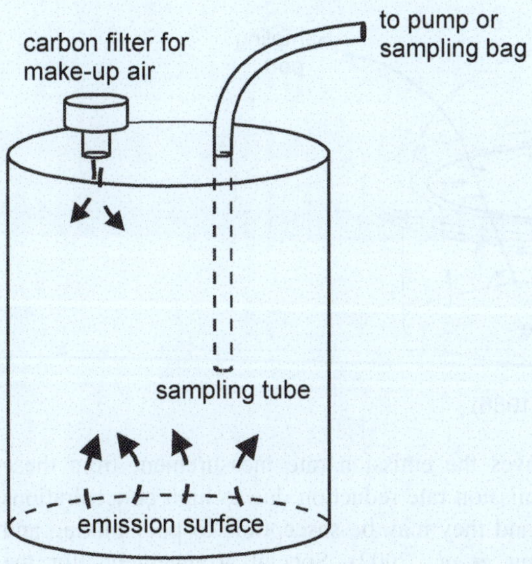

Figure 3.7. Equilibrium chamber (after Bowker *et al.*, 1994).

Due to the very low velocities in these chambers, there is no enhancement of emission rate by reduction of boundary layers. In addition, emission rates will reduce over time, approaching zero as equilibrium concentrations are reached, assuming the sample airflow is small compared to the chamber volume. These chambers are likely to predict very low emission rates as a result. There are several examples of this kind of chamber being used for emission rate sampling in the literature. Hodgson and Paspaliaris (1996) used a static flux chamber to measure odour emissions from wastewater lagoons. Tillman *et al.* (2002) compared static flux chamber results against modelled results for a soil test-cell. Measured and modelled fluxes were found to be in good agreement for diffusive and diffusive-convective emissions. This flux chamber utilised recirculation of sampling air, and employed adsorbent traps for collection of samples, so it would have had relatively good mixing compared to a completely static chamber and may also have been less susceptible to suppression of emissions due to concentration build-up in the chamber. Perera *et al.* (2002) also performed tests on a static flux chamber on a soil test-cell. Emission rates measured with the flux chamber were found to underestimate actual flux rates by 15–65% due to increased concentrations within the chamber. The largest errors occurred for the largest fluxes and smallest chamber volumes. Mosher *et al.* (1999) applied static flux chambers to measurements of landfill gas emissions and found these to be in good agreement with micrometeorological measurements. Care was taken to minimise disruption to the sampled surface, and short measurement times were employed. This would act to minimise emission suppression through the build-up of high concentrations in the chamber.

Dynamic flux chambers

Dynamic flux chambers are similar to static flux chambers but utilise a measured airflow as part of the emission rate measurement. The airflow used in dynamic flux chambers is small, typically in the range 1–40 L min^{-1}. The emission rate is determined by the product of the airflow through the chamber and the odour concentration. An example flux chamber is shown in Figure 3.8.

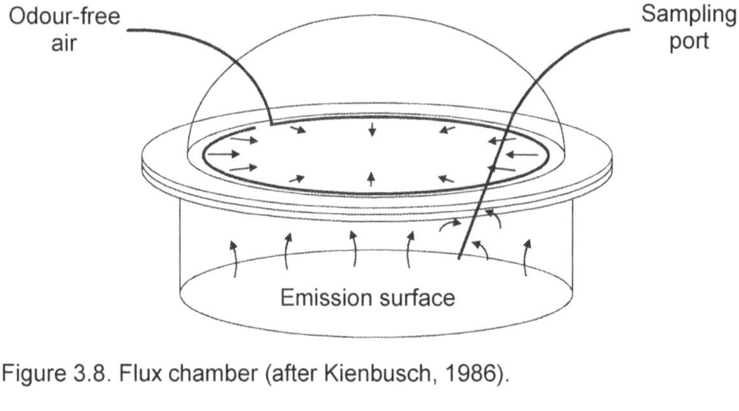

Figure 3.8. Flux chamber (after Kienbusch, 1986).

The incorporation of an airflow improves the emission rate measurement from these devices, but they are still susceptible to emission rate reduction due to high concentrations within the chamber if the airflow is low, and they may be susceptible to poor mixing and dead zones (Gao and Yates, 1998; Zhang *et al.,* 2002). Special arrangements for the introduction of the sweep flow into the chamber, such as circular flow rings, are intended to promote good mixing (Cooper *et al.*, 1992). Circulation fans may also be included within the chamber to improve mixing. It is also important to allow steady-state conditions to be achieved. Initial concentrations in the chamber may be high if it is placed on the emission surface prior to allowing the sweep air to flow through; this could produce an artificially high emission rate measurement if not allowed to stabilise (Park and Shin, 2001).

As emissions have been shown to increase with airflow in some cases in dynamic flux chambers (Bowker *et al.,* 1994; Cooper *et al.,* 1992; Lindberg *et al.,* 2002; Sadek *et al.,* 1998) there is still the problem of selecting a suitable airflow for the emission measurement. Dynamic flux chambers such as that shown in Figure 3.8 pose problems as it is not possible to determine the flow velocity in the chamber; this makes it difficult to relate the flux chamber measurement to field conditions where wind speed may be an important determinant on emission rate.

There are several examples of the use of dynamic flux chambers in the literature, for both liquid and soil emissions. Adams *et al.* (1981) used dynamic flux chambers to determine biogenic sulphur emissions from soils. Park and Shin (2001) and Guy and Page (2001) applied dynamic flux chambers to landfill gas emissions. Frechen (1992) applied a flux chamber to measurements of odour emissions from sewage treatment processes. The Frechen flux chamber differed from the type shown in Figure 3.8, in that it was rectangular, allowing flow velocities to be calculated. Flow rates of 9–34 L min^{-1} were used, equating to an air sweep velocity of 2.5–9.5 mm s^{-1}.

Wind tunnels

Wind tunnels, like dynamic flux chambers, utilise an airflow in the emission rate measurement. Unlike many flux chambers, wind tunnels have a prismatic measurement section, with the airflow being blown longitudinally along the tunnel. This enables the sweep velocity to be controlled more precisely, allowing the variation in emission with velocity to be studied, and wind-tunnel measurements to be related to field conditions. Airflows are very much higher than for dynamic flux chambers, with sweep velocities of 0.1–5 m s^{-1} being common.

There are numerous examples of portable wind tunnels which have been used for emission rate measurement in the literature. Several examples are shown in Figure 3.9.

One of the earliest examples of a portable wind tunnel being applied to the collection of odour samples was described by Lindvall *et al.* (1974). The tunnel consisted of a rectangular measurement section, open to the emission surface, with contraction and expansion sections allowing for air to be blown through the measurement section. Figure 3.10 shows a Lindvall type hood being used for odour emission rate measurements on a landfill site. Another early wind tunnel system was described by Lockyer (1984) which was applied to ammonia volatilisation from grass swards. The aim of this system was to allow ammonia emission (or uptake) measurements from growing plants in the field without influencing natural growing conditions. The measurement section of the wind tunnel system was constructed of transparent polycarbonate to allow sunlight penetration. The air velocity in the wind tunnel was matched to the wind velocity at 25 cm outside the tunnel, allowing the growing plants to be exposed to realistic wind velocities. Temperature differences inside and outside the tunnel were found to be small. Condensation problems were apparent on the walls of the tunnel, which was eventually prevented by using car windscreen type heating panels. Braschkat *et al*. (1993) describes a large wind tunnel system which was applied to ammonia emission from soils by Mannheim *et al*. (1995). As in Lockyer's (1984) experiments, air velocity in the wind tunnel was matched to ambient wind speeds. A similar wind tunnel was applied to ammonia emissions from soils by Reitz and Kutzbach (1997), and a tunnel similar to the Lindvall *et al*. (1974) hood was applied to odour and hydrogen sulphide emissions from manure by Schmidt and Bicudo (2000).

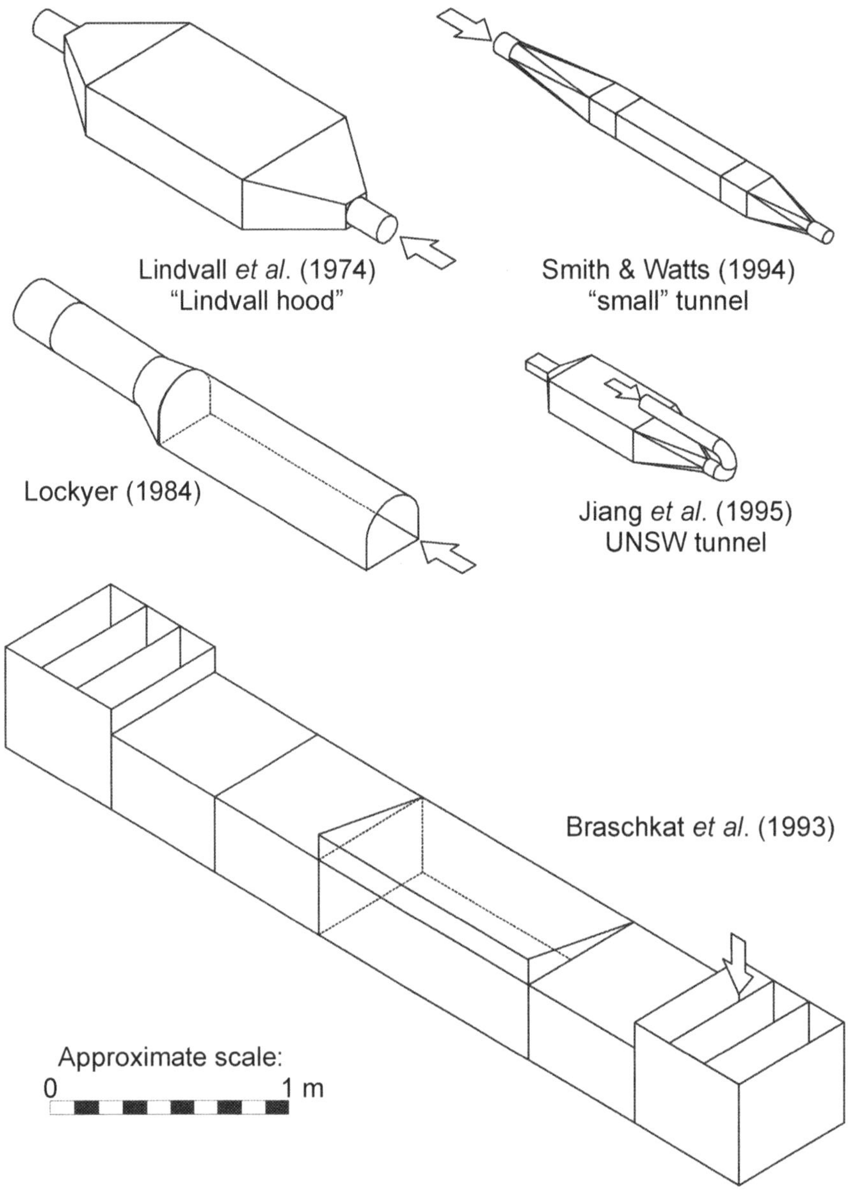

Figure 3.9. Examples of wind tunnels used for emission measurement.

Figure 3.10. Lindvall type hood being used for odour emission rate measurement at a landfill site (courtesy of Silsoe Research Institute, UK).

For a wind tunnel system to be useful it is necessary to have an understanding of the aerodynamics of the tunnel. It is important that concentration and air velocity measurements are representative of the conditions at the emission surface and can be related to field conditions. Early examples of wind-tunnel systems have been found to have highly non-uniform velocity and concentration profiles which can lead to errors in emission rate estimation. For example, van Belois and Anzion (1992) experimented with acetone emission rates measurement using two wind tunnel systems, these being similar to the Lindvall *et al.* (1974) hood and the Lockyer (1984) wind tunnel. Velocity and concentration profiles within the hoods were found to be non-uniform, leading to complications in obtaining representative velocity and concentration measurements. Similarly, Smith and Watts (1994b) experimented on 'large' and 'small' wind tunnels. The large tunnel was similar to the tunnel of Lockyer (1984) and the small tunnel is as shown in Figure 3.9. The tunnels were applied to ammonia emissions from feedlot pads. Results showed that the ammonia emission rate variation with wind velocity was similar for each tunnel, but differences were apparent in the emission rates, with the smaller of the two wind tunnels consistently giving a higher emission rate. This was thought to be due to different velocity profiles within the wind tunnels, the smaller tunnel having a higher velocity close to the emission surface.

Jiang *et al.* (1995) and Bliss *et al.* (1995) have developed the UNSW wind tunnel from the original Lindvall (1974) hood. Experimentation with the Lindvall hood showed the velocity profile in the measurement section to be variable and non-uniform. A much improved velocity profile was achieved by extending the inlet duct, and fitting vanes and a baffle in the expansion section. Velocity profiles for the UNSW wind tunnel are shown in Figure 3.11, and an example of the hood being used for emission rate measurement from a lagoon is shown in Figure 3.12.

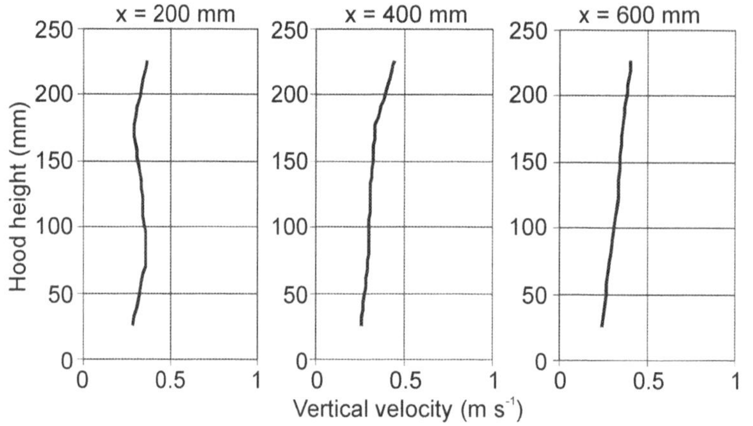

Figure 3.11. Velocity profiles in the UNSW wind tunnel (after Jiang *et al.,* 1995).

Figure 3.12. UNSW wind tunnel being applied to emission rate measurements from a lagoon (courtesy of University of New South Wales).

Selection of suitable wind speeds for wind tunnels can be difficult. Lockyer (1984) and Mannheim *et al.* (1995) varied wind-tunnel air velocities to match ambient wind speeds during the course of measurement. Jiang and Kaye (1997) determined the appropriate velocity for the UNSW wind tunnel on the basis of wind speeds associated with the most odour complaints. It is advantageous to use a range of velocities, and derive the relationship of emission rate with wind velocity. This allows the use of variable emission rates in subsequent modelling, which can have significant results on model output. Wind-tunnel derived emission rates are generally taken as a function of the average velocity across the tunnel section. It is more useful if the emission rate is taken as a function of the velocity profile (Smith and Watts, 1994b; Leyris *et al.*, 2000), which could be expressed in terms of shear stress, velocity profile or friction velocity. This allows wind-tunnel measured emission rates to be adjusted for different wind speeds using theoretical and empirical emission models. Smith and Watts (1994b) and Leyris (2000) derived friction velocities from wind-tunnel velocity profiles.

It is important to take into account that concentration profiles within wind tunnels are distinctly non-linear, as shown in Figure 3.13. This has implications for the withdrawal of odour samples from the wind tunnel. If single-point sampling is used, it is necessary to ensure that the airflow is well mixed, which is the purpose of the contraction section in many wind-tunnel designs. Contractions may, however, have the disadvantage of increasing pressures within wind tunnels, leading to suppression of convective emissions on porous surfaces. Multi-point sampling is also an option, whereby samples are collected from several positions from the flow cross-section, and a spatial average obtained. This has been utilised in tunnels without mixing chambers (Loubet *et al.,* 1999a) or in conjunction with mixing chambers (Wang and Kaye, 2001).

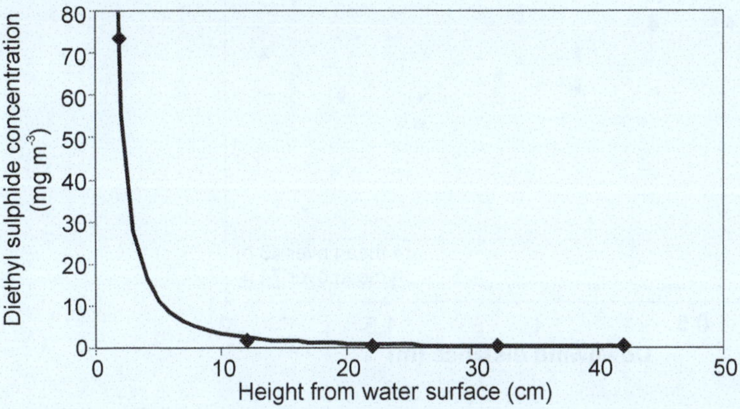

Figure 3.13. Concentration profiles in wind tunnel (after Leyris *et al.,* 2000).

The length of the emission sample area (the fetch) can also influence results. It would be expected that emission rates at the leading edge of the sample area would be greater than at the trailing edge. This is because gas concentration in the boundary layer increases with fetch, in effect reducing the driving force for mass transfer. Loubet *et al.* (1999a, 1999b) investigated fetch effects on a tunnel similar to the Lockyer (1984) tunnel and found that a stable boundary layer did not develop until approximately 100–150 cm within the measurement section, after which emission measurements become independent of fetch. This implies that longer wind tunnel sections would be desirable, although this can limit the practicality of the tunnel. It should be possible to compensate for fetch effects using mass-transfer modelling. Figure 3.14 illustrates fetch effects.

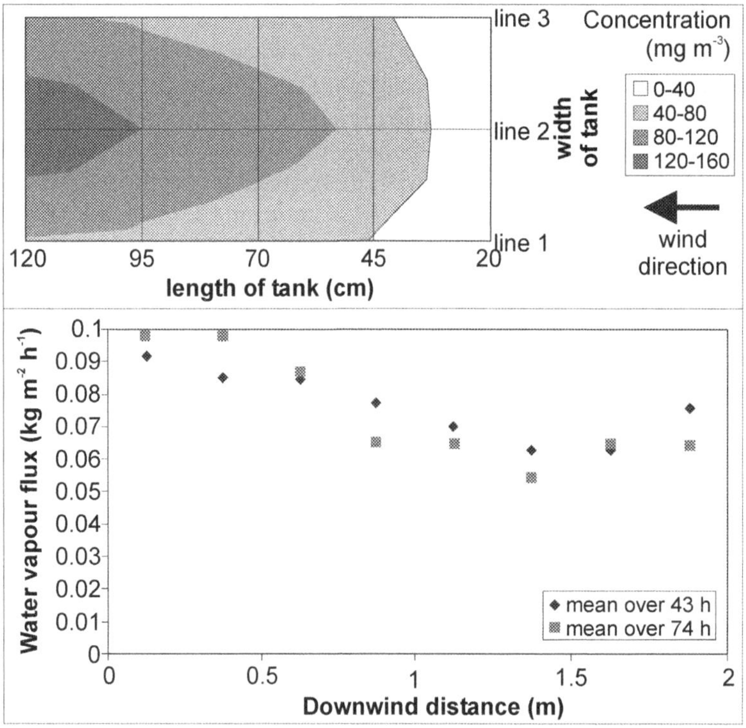

Figure 3.14. Fetch effects on emission rate. Top: diethyl sulphide concentrations measured 2 cm above water surface in a wind tunnel (Leyris, 2000). Bottom: reduced evaporation rate with fetch as measured by Loubet *et al.* (1999b).

A wind tunnel system, as well as having well defined aerodynamic properties, should also be practical in use. For measurement of emissions from liquid surfaces, wind tunnel systems are typically fitted with floats or suspended above the liquid surface using ropes. The area covered by the hood is a trade-off between resolution and practicality – higher resolutions require larger areas and make the hood system increasingly unwieldy. It is necessary to be able to provide a supply of odour-free air to the inlet-end of the tunnel. For systems with an inlet expansion section, such as the Lindvall hood or UNSW wind tunnel, this is a relatively easy undertaking. It is more difficult for systems such as the Lockyer wind tunnel, since measurements both upwind and downwind of the emission surface are required. If olfactometry is used for odour measurement, the inlet concentration is likely to be low and difficult to measure accurately. The lack of an inlet expansion section also makes the Lockyer hood susceptible to interference by external winds (Loubet *et al.*, 1999b).

Small tunnels of the Lindvall type appear to offer the best compromise between practicality and accuracy. The UNSW tunnel represents a useful development of the Lindvall hood, and would appear to be the hood of choice for emission measurement from liquid surfaces. Useful work has been performed on defining the aerodynamics of the tunnel (Jiang *et al.*, 1995), although further work is required to determine velocity profiles near the liquid surface and to determine friction/bulk velocity correlations.

Several studies have been performed on the accuracy of wind tunnel systems. Reitz and Kutzbach (1997) and van der Weerden *et al.* (1996) analysed ammonia recovery rates for wind tunnel systems and found these to be in the range 75–100%. Bliss *et al.* (1995)

compared the performance of the UNSW wind tunnel against a boundary layer model for the emission of ammonia from a water surface. The wind tunnel and the model were in good agreement for air velocities less than about 0.6 m s^{-1}, but the wind tunnel emission rates were consistently about 10% lower than the model predictions; this was attributed to bias in the ammonia concentration measurement method. For air velocities greater than 0.6 m s^{-1}, the wind tunnel emission rates were higher than the model predictions, this being due to wave formation, which in turn led to droplet formation and accumulation on the surface of the hood. Mannheim *et al.* (1995) compared tunnel ammonia emission rates to emission rates derived by the integrated horizontal flux method. It was concluded that tunnel and micrometeorological measurements were similar, with the tunnel being more accurate and also applicable to lower wind velocities than the integrated horizontal flux method.

3.4.3 Choice of area source measurement method

The discussion above has shown that the area source emission rate is not straightforward. Although several of the techniques are simple in principle, they have the potential to influence emission rates and require great care in their application.

Many area sources may vary substantially in emission rates, both spatially and temporally. Sewage treatment works, for example, may show considerable temporal variations depending on influent odorant concentrations, flow rates, and for certain processes, meteorological variations, such as wind speed and air temperature. Landfill sites can also have considerable temporal and spatial variations. Temporal variations can arise from the different phases of gas generation as the filled waste degrades (Cooper *et al.*, 1992). Shorter-term variations may result from changes in air pressure at the landfill surface, caused by weather patterns, temperature-related diurnal pressure variations, or more rapid fluctuations due to the action of the wind blowing over the surface (Stern *et al.,* 1999). Temporal variations can also result from changes in soil moisture due to precipitation. Considerable temporal and spatial variations may be observed without any obvious relationship to these variables (Milton *et al.,* 1997). Spatial variations on landfill sites can result from different cells being at different stages of decay, heterogeneous wastes, heterogeneous cover materials, the presence of cracks and fissures (Cooper *et al.,* 1992). Sampling programmes should be designed so that they are adequate to capture the variability in emission rates.

Where emission rates vary spatially, employing a micrometeorological technique can be advantageous as these can measure the total emissions from a large site. Unfortunately, micrometeorological techniques cannot be applied to all sites, particularly where emission rates are very low, or where there are upwind odour sources which may interfere with concentration measurements. For hood techniques, as many locations in as short a time period as possible should be sampled in order that the hood measurements can be reliably extrapolated to give site-wide flux values (Mosher *et al.,* 1999). This means the hood technique used must be very portable and capable of making rapid flux measurements. Mosher *et al.* (1999) found static flux chambers to be useful for this purpose: a team of three individuals were able to conduct 60–100 flux measurements per day on a landfill site.

For emission surfaces where diffusive mass transfer is important, the influence of wind speed on emission rates has to be considered. This is particularly important for liquid surfaces because of the influence of wind on both the air-side boundary layer and the increased turbulence on the liquid side due to the formation of waves. Waves also serve to increase the interfacial area for mass transfer. The action of the wind is significant for many soil surfaces, even when there is a significant convective component such as may occur for landfill sites. Investigators have shown relationships between wind speed and emission rates for landfill sites, resulting from reductions in boundary layer thickness and increasing concentration gradients (Cooper *et al.*, 1992; Milton *et al.*, 1997).

Because emission rates can vary with wind speeds, it is advantageous to develop a relationship between wind speed and emission rate, especially for liquid surfaces. Indeed, if emission rates are being measured to serve as the input to dispersion models, it is essential that the relationship with wind speed be included if the model results are to be meaningful. This would suggest that wind-tunnel type hoods are the best method, as these are the only hood method that allows simulation of realistic wind speeds over the emission surface. A number of issues, however, require further research before reliable wind tunnel measurements can be made; these include the relationship between wind profiles in the hood and in the field, concentration profiles, fetch effects and pressure effects.

A major disadvantage of wind-tunnel type hoods arises from the relatively large airflows used. These result in dilution of the emitted gases, so that wind tunnel methods can lack sensitivity and be unsuitable for very low emission rates. This is an area where static and dynamic flux chambers can be preferable, as the relatively low sweep airflow means there is minimal dilution of the emitted gases (Perera *et al.,* 2002). For circumstances where emission rates are low and wind effects are minimal, such as when emission surfaces are shielded from the effects of the wind by vegetation, static or dynamic flux chambers can be a good method, provded that care is taken to minimise concentration gradient or pressure effects by using short sampling times (Cooper *et al.,* 1992, Mosher *et al.,* 1999).

The odour measurement technique employed also has an influence on the area source measurement technique used. If olfactometry is used, practical limitations of the test procedure mean the number of samples will be limited. In addition, the sensitivity of olfactometry is relatively poor, so high odour concentrations in the samples are preferable. This can pose a particular problem for spatially variable sites. The requirement for a low number of samples suggests a whole-site micrometeorological method is preferable, but the requirement for relatively high concentrations suggests hood methods would be better. Unfortunately, hood methods require a large number of measurements to capture adequately the spatial variability. The choice of emission measurement method will often be the result of such trade-offs, and it is not possible to recommend a universally suitable method.

The advantages, disadvantages and potential applications of the different area source measurement methods are listed in Tables 3.1 and 3.2.

Table 3.1. Advantages, disadvantages and potential applications of micrometeorological area source emission rate measurement methods.

Technique	Advantages	Disadvantages	Applications
All	Can measure whole site emissions. Passive measurement – does not interfere with emission mechanisms. Not limited to static surfaces.	Measures dispersed odours – potential low odour concentrations and poor sensitivity. Sensitive to upwind interference.	Isolated solid or liquid sources without upwind interference. Sites with static or dynamic surfaces.
Flux profile methods	Not sensitive to site geometry. Potentially little modelling input required.	Large number of measurements required.	Heterogeneous sites where whole-site emission rate is to be determined. Emissions from solids or liquids
OP-FTIR	Measures average concentration along line up to 1 km long. Measurements are rapid, once equipment is in place. Can potentially be combined with tomography to measure spatial variations.	Large equipment requirement. Can be unsuitable for sites with limited access or complex topography. Limited range of analytical odour concentration measurements only.	Heterogeneous sites where whole-site emission rate is to be determined, provided good access and simple topography. Heterogeneous sources where spatial variations are to be studied (combined with tomography). Odorant emissions from solids or liquids.
Integrated horizontal flux method, theoretical profile shape method	Reduced number of measurements required.	Limited to strip or radial sources. Increased reliance on modelling for velocity and concentration profiles.	Solid or liquid surface emissions from line sources, radial test plots, possibly radial sedimentation tanks.
Dispersion model methods	Reduced number of measurements required. Flexible in terms of measurement locations.	Total reliance on models for velocity and concentration profiles.	Heterogeneous sites where whole-site emission rate is to be determined. Emissions from solids or liquids.

Table 3.2. Advantages, disadvantages and potential applications of hood area source emission rate measurement methods.

Technique	Advantages	Disadvantages	Applications
All	Isolates portion of emission surface, so can be used on complex sites with upwind interference. Higher concentrations measured – potentially more sensitive.	Many measurements required for heterogeneous sources. Potential to interfere with emission mechanisms; Can be difficult to relate conditions in the hood to field conditions. Suited to static surfaces only. Can be difficult to form effective seal on some surfaces.	Complex sites with upwind interferences. Detailed surveys of different emission sources on a site.
Static flux chambers	Little dilution of emissions – good for low emission rates; Low equipment requirements. Rapid measurements.	Diffusive emissions can be suppressed through high chamber concentrations. Poor representation of boundary layer. Poor mixing in the chamber. Not suited to emissions from liquid surfaces.	Useful for rapid measurements at many locations. Emissions from heterogeneous sites where spatial variability is being studied. Emissions from sheltered solid surfaces where wind effects are negligible.
Dynamic flux chambers	Potentially greater control over measured concentrations by varying sweep airflows.	Convective emissions can be misrepresented due to pressure effects. Diffusive emissions can be suppressed through poor representation of boundary layer; Can be slow to stabilise. Greater equipment requirements. Not suited to emissions from liquid surfaces.	Emissions from relatively homogeneous sheltered solid surfaces where wind effects are negligible.
Wind tunnels	Greater control over variables influencing emissions. More accurate representation of wind effects. Potential to develop wind/ emission rate relationships for dispersion model input.	Large equipment requirements; Difficulties in selecting/measuring representative wind speeds. Care required in design, particularly in terms of velocity and concentration profiles and pressure effects.	Emissions from relatively homogeneous solid or liquid surfaces where wind effects are significant.

4
Sample collection and storage

4.1 INTRODUCTION

Because olfactometry and several analytical measurements cannot practically be performed on site, odour samples need to be collected, preserved and stored to allow transportation from site to laboratory. This is an area that requires careful attention if a representative measurement is to be made.

There are several 'trapping' methods in use for sample collection for gas-chromatography, which selectively trap and thus pre-concentrate odorants for subsequent laboratory analysis. This is a specialised area with application to many gaseous analyses as well as odorants, and detailed discussion is beyond the scope of this report. For further information, Hobbs (2001) provides a useful review of trapping methods for GC analysis.

Sample collection for olfactometry tends to employ containers which are filled with the odour sample, sealed and then transported to the laboratory. Generally, flexible plastic bags are used, although other methods such as stainless steel canisters are available, usually with a significant weight and cost penalty. Typically, bags are filled using the 'lung' method, whereby the sample is drawn into them by subjecting the outside of the bag to negative pressure. Sampling by the lung-method is illustrated in Figures 4.1 and 4.2.

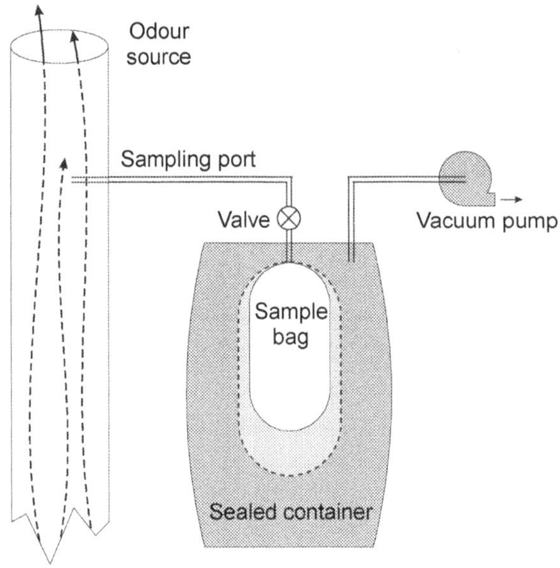

Figure 4.1. Bag sampling using lung method.

It is important that the sample is not contaminated, lost or altered during storage. This leads to several desirable properties for sample bags. Odour bags should be (Jiang and Kaye, 2001):

- odour-free, i.e. they should not add odours to the sample;
- of a material which does not adsorb or react with odorous samples;
- sufficiently impervious to prevent any significant loss of odour components;
- reasonably robust;
- leak-free;
- equipped with leak-free fittings, compatible with olfactometer and other sampling equipment;
- of sufficient capacity to enable a full test series to be completed.

Figure 4.2. Odour sample bag and vacuum container for lung method of sample collection (Courtesy of IGER).

4.2 MATERIALS

Bag materials are obviously an important factor in sample preservation. The material should be odour-free and impermeable, and should not encourage adsorption of odorants. It should also be flexible and reasonably durable. Several materials have been used, including tedlar (polyvinyl fluoride), teflon (polytetrafluoroethylene), nalophan, polyterephthalic ester co-polymer, mylar, saran and polyethylene. Several studies (Andino and Butler, 1991; Fan *et al.,* 2001; Fugit *et al.,* 1996; McGarvey and Shorten, 2000; Pau *et al.,* 1991; Pet'ka *et al.,* 2000; Schuetzle *et al.,* 1975; Sulyok *et al.,* 2001; Wang *et al.,* 1996) have been carried out on different bag materials, and have generally concluded that tedlar bags are superior to other materials and can in many cases preserve samples without significant losses for several days. Figure 4.3 shows ethylbenzene losses from several bag materials over a period of 200 minutes. This indicates that tedlar is clearly the best material, with small losses compared to the rapid losses for polyethylene, saran and mylar. Figure 4.4 shows losses of several reduced sulphur compounds in tedlar bags over a period of 400 hours. Losses are in general small, maximum losses of about 20% occurring for methyl mercaptan after 300 hours.

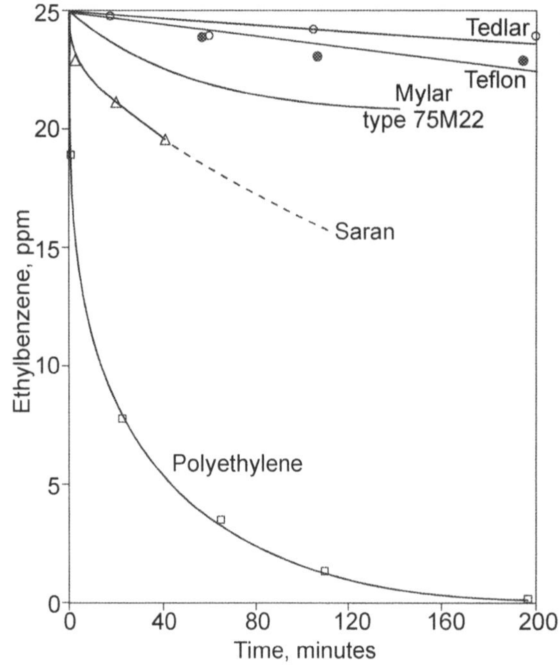

Figure 4.3. Stability of ethylbenzene in different bag baterials (after Schuetzle *et al.,* 1975).

Adsorption losses are known to occur in tedlar bags, particularly for the more volatile polar organics (McGarvey and Shorten, 2000; Pau *et al.,* 1991). Because the most polar compounds tend to be the most odorous compounds, this has serious implications for preservation of odour samples (Schuetzle *et al.,* 1975). For this reason, it is usually advised that samples are measured as soon as possible after collection, preferably within 24 hours (McGarvey and Shorten, 2000; Sulyok *et al.,* 2001). It would also appear advantageous to use larger volume bags where possible, as this reduces the surface-area/volume ratio (Andino and Butler, 1991).

Care should also be taken to prevent contamination and adsorption in sampling lines and fittings. Provision should be made for lean sample lines to be used for each sample (Jiang and Kaye, 2001). Materials such as teflon, stainless steel, glass etc. are often recommended for sample lines and fittings. Teflon is perhaps the best choice, as indicated by studies on tedlar bag fittings studies by Wang *et al.* (1996) who found significant VOC losses to occur where stainless steel fittings were used. The losses were greatly reduced with teflon fittings.

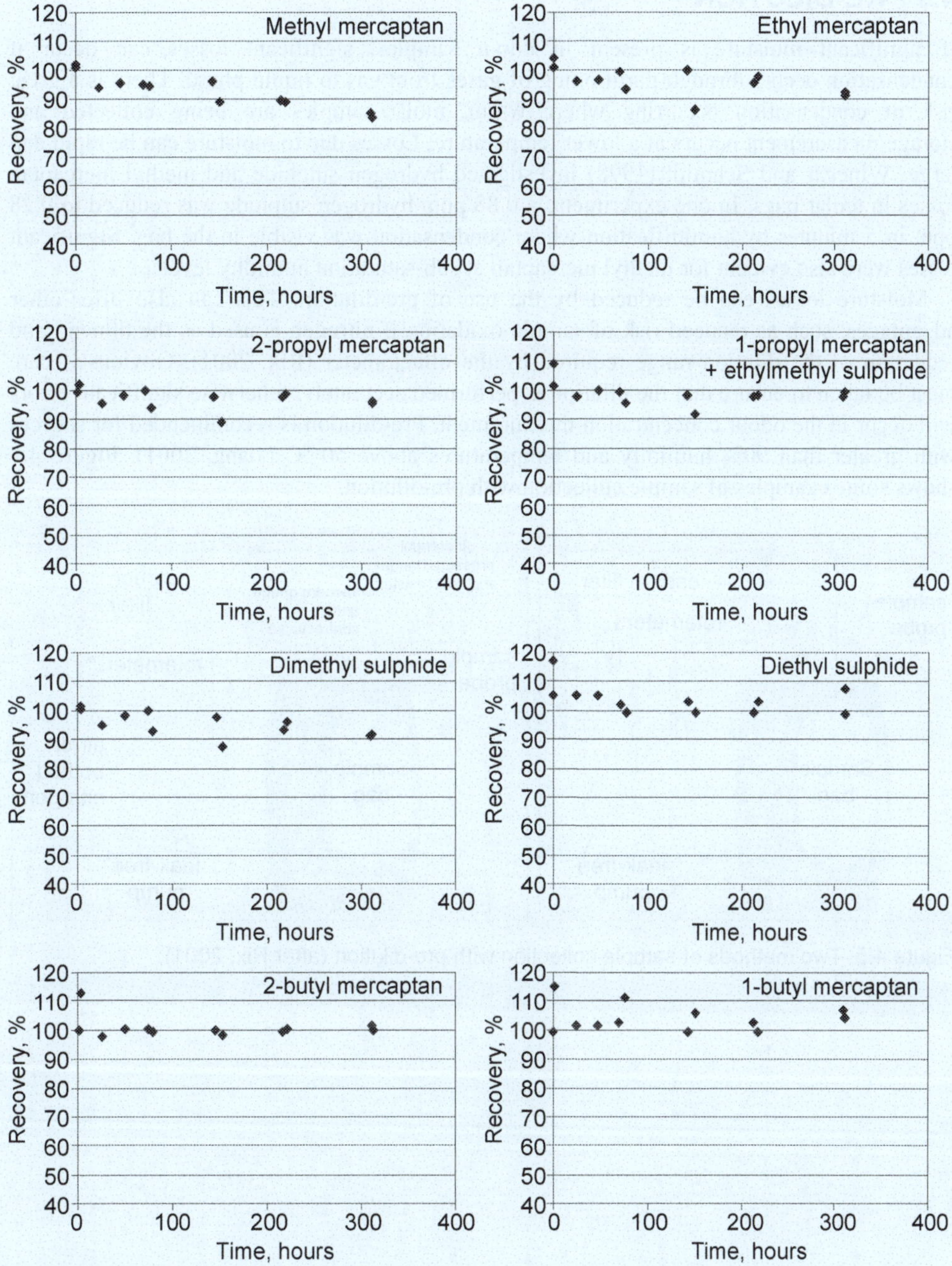

Figure 4.4. Recoveries of reduced sulphur gases from tedlar bags (after Sulyok *et al.*, 2001).

4.3 PRE-DILUTION

If significant moisture is present in odour samples, significant losses can occur if condensation occurs through partitioning of gases from gas to liquid phase. There is a great risk of condensation occurring where warm, moist samples are being collected and storage/measurement occurs at a lower temperature. Losses due to moisture can be rapid and large. Winegar and Schmidt (1998) investigated hydrogen sulphide and methyl mercaptan losses in tedlar bags. In one experiment, a 0.88 ppm hydrogen sulphide was reduced to 0.28 ppm in 5 minutes by humidification where condensation was visible in the bag. Significant losses were also evident for methyl mercaptan at sub-saturation humidity levels.

Moisture losses can be reduced by the use of pre-dilution. This can also offer other advantages, such as reduced risk of sample oxidation if nitrogen is used as the diluent, and reduction of the dilution range required by the olfactometer (Rix, 2001). Obviously, care must be taken to ensure that the dilution is performed accurately, otherwise significant errors will occur in the odour concentration measurement. Pre-dilution is recommended for sources with greater than 90% humidity and temperatures above 50 °C (Jiang, 2001). Figure 4.5 shows some examples of sample collection with pre-dilution.

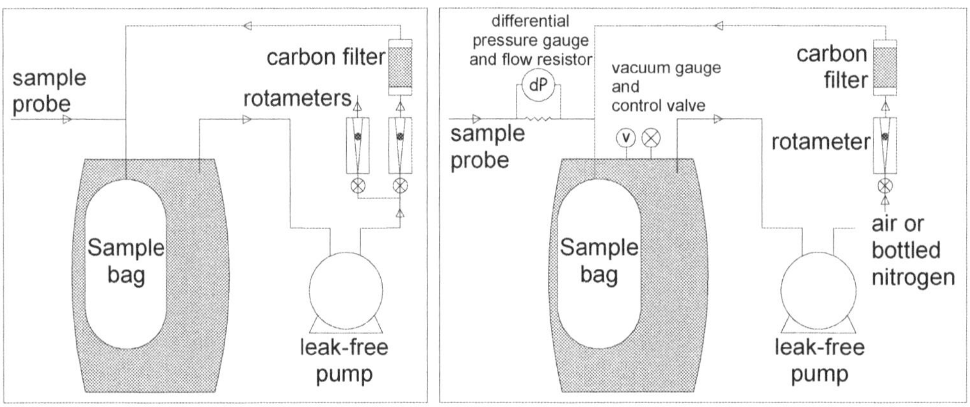

Figure 4.5. Two methods of sample collection with pre-dilution (after Rix, 2001).

5

Conclusions

5.1 ODOUR MEASUREMENT

Odour measurement is a difficult area. Because the linkage between odorants and odours is unclear, there is no single analytical measurement which can be related to odour perception, as there is for noise. Sensory measurements are in common use to classify odour sources.

Threshold olfactometry has emerged as the sensory method of choice for nuisance odour assessment, principally due to recent European efforts to improve repeatability and accuracy through standardisation of the test procedure. Olfactometry conforming to CEN draft standard prEN 13725 is commonly regarded as best practice for sensory odour measurements.

Standardisation has undoubtedly improved threshold olfactometry, but it will never achieve the levels of precision associated with analytical measurements. It is important to consider the level of uncertainty involved with olfactometry, requiring multiple samples to be taken if this is to be reduced. Triplicate or quadruple samples are recommended as a compromise between cost and precision. More samples may be required if odour concentrations are low (< 1000 ou m^{-3}), due to the poor sensitivity of olfactometry.

Olfactometry has a number of other disadvantages.

- Olfactometry is a laboratory-based measurement, making sample storage and transportation necessary.
- Olfactometry requires a panel of several observers and a series of dilutions to obtain a concentration measurement. This makes measurements slow and expensive.
- Olfactometry suffers from poor sensitivity and is unsuitable for background odour levels (< 50 ou m^{-3}).
- The odour concentration measurement provides no information on the composition of the odour.

© IWA Publishing. *Sampling for the Measurement of Odours*. P. Gostelow, P. Longhurst, S. A. Parsons and R. M. Stuetz. ISBN: 1 84339 033 7

Olfactometric measurements relate to human odour perception and are, therefore, well suited to assessing annoyance at odour receptors. Olfactometric measurements are often applied to emission rate measurements for the purpose of dispersion modelling. They may be less appropriate for this purpose. Different odorants will transfer from solid or liquid phase to gas phase in different ways, depending on their physical and chemical characteristics. They may be chemically transformed in the gas phase, changing the character of an odour between source and receptor. Where different odour sources are present in a single site, the different odorants present may have synergistic or antagonistic effects on the overall odour when they combine in a plume downwind of the site. These effects are impossible to predict using atmospheric dispersion modelling of odour concentrations and for practical purposes are usually ignored.

Analytical measurements refer to odorants and are therefore more suited to source measurements. Consideration of odorants, rather than odours, allows modelling of physical and chemical mechanisms of emission and transformation. Unfortunately, analytical measurements alone tell us little about the sensory effect of the odorants on humans.

It is not possible to indicate best practice for analytical measurements. There is no universal marker compound that can be easily related to odours as perceived. Hydrogen sulphide has some advantages and has proven to be useful for some applications, specifically for sewage and some agricultural situations, but its usefulness as a marker for waste management processes is at present unknown.

It is clear that linkages between analytical and sensory measurements are required. This is a difficult area because many environmental odours consist of many odorants and the effect of odorant mixtures on the resulting odour is difficult to predict. However, recent work such as Gostelow and Parsons (2000), Hobbs *et al.* (2001) and Noble *et al.* (2001) has shown that odours can be predicted from a few key odorants in certain environmental odours. It would obviously be beneficial if this research could be extended to other environmental odours and this should be considered a major research priority.

5.2 SOURCE SAMPLING

When sampling odour sources, the emission rate is of vital importance. The ease of emission rate measurement depends largely on the source configuration. Point sources are relatively easy, volume sources may be easy if the emission points are controlled or difficult if they are uncontrolled, and area sources are almost always difficult due to lack of a controlled airflow.

Micrometeorological methods are currently best suited to relatively large, homogeneous sources, especially where topography is uncomplicated. Of the methods available, emission rate estimation by dispersion modelling offers several advantages, in particular the relative flexibility regarding sampling locations and source configurations. The availability and continuous development of dispersion modelling packages can simplify micrometeorological emission rate estimation and extend its application to less homogeneous sites.

The range of source types and ease of measurement could be further improved through further research in two areas. First, OP-FTIR methods could simplify micrometeorological measurements through the provision of averaged concentration measurements over long path lengths. The use of CAT in conjunction with OP-FTIR is also an exciting development in this field, providing two-dimensional concentration maps of relatively large and heterogeneous sources within relatively short periods of time. The second area is further development of dispersion models for emission rate prediction. Specific areas for consideration are the handling of complex topography and short averaging times. Application of improved dispersion models should allow for emission rate measurement from more complex sources.

Hood methods can be advantageous as they offer a more controlled environment for measuring surface emission rates. For many emission sources, the wind blowing over the surface can be a major influence on the emission rate. This is particularly true for liquid surfaces. For these source types, the wind tunnel type hood is the method of choice. Hoods with low or poorly defined airflows, such as static or dynamic flux chambers, are not recommended for sources where wind effects are significant. They can, however, be used for sheltered emission surfaces and have the advantage of higher odour concentrations and potentially greater sensitivity. Static flux chambers in particular are useful for covering many measurement locations in a short period of time. Care must be taken to prevent very high concentrations occurring in the chambers, however, as this can suppress emissions.

Although wind-tunnel type hoods have undergone development over recent years, there are a number of areas for further research. Key amongst these is the ability to relate measured wind tunnel emission rates to field emission rates over a range of meteorological conditions. This is best achieved by careful attention to velocity profiles within the wind tunnel and their relationship to velocity profiles in the field. The use of parameters such as the friction velocity is useful, as it allows measured emission rates to be adjusted using mass-transfer modelling. Other areas for improvement include consideration of concentration profiles, with implications for sample collection points and length (fetch) of tunnels, pressure effects, choice of materials, and the trade-off between minimum enclosed area and practicality. It is clear that there are still many unknowns with wind-tunnel sampling systems and they could be greatly improved.

5.3 SAMPLE COLLECTION AND STORAGE

Sample collection and storage are required for olfactometry. Bag sampling is commonly used for reasons of performance, cost and practicality. Care should be taken in the choice of bag and fitting materials. Tedlar bags with teflon fittings appear to be the best option, and larger bags are preferred as they reduce the surface area/volume ratio.

Sample losses over time are inevitable in sample bags, especially for the more volatile polar odorants. Although some experiments have been carried out on sample losses, it is difficult to generalise these results and arrive at specific recommendations, other than to reduce the time delay between sample collection and measurement as much as possible.

Humidity can potentially have a great effect on odour sample degradation, but there is little information in the literature other than the recommendation that moisture within bags should be prevented. Pre-dilution is commonly used to prevent condensation for warm, moist samples, and is recommended for those with relative humidity > 90% and temperatures above 50 °C.

Because of the potential for sample losses, and the largely unquantified effects of humidity on bag samples, this area clearly needs further research.

6

References

Adams, D.F., Farwell, S.O., Pack, M.R. and Robinson, E. (1981) Biogenic sulfur gas emissions from soils in eastern and southeastern United States. *Journal of the Air Pollution Control Association* 31, 1083–1089.

Andino, J.M and Butler, J.W. (1991) A study of the stability of methanol-fueled vehicle emissions in Tedlar bags. *Environmental Science and Technology* **25**(9), 1644–1646.

Anon. (1998) Agency acts on odour problems at Biffa and 3C landfills. *Ends Report* **281**, 14–15.

Bird, R.B, Stewart, W.E, and Lightfoot, E.N. (2002) *Transport phenomena* (2nd edn). John Wiley & Sons, New York.

Bliss, P.J, Jiang, K. and Schulz, T.J. (1995) The development of a sampling system for the determination of odor emission rates from areal sources: Part II. Mathematical model. *Journal of the Air and Waste Management Association* **45**, 989–994.

Bliss, P.J., Schulz, T.J., Senger, T. and Kaye, R.B. (1996) Odour measurement: factors affecting olfactometry panel measurement. *Water Science and Technology* **34**(3–4), 549–556.

Bowker, R.P.G., McGinley, C.M. and Webster, N.A. (1994) Estimating odour emission rates from wastewater treatment plants. Water Environment Federation Specialty Conference: *Odors and Volatile Organic Compound Emission Control for Municipal and Industrial Wastewater Treatment Facilities.* Jacksonville, FL, 24–27 April 1994, pp. 3.13–3.24.

Braschkat, J., Mannheim, T., Horlacher, D. and Marschner, H. (1993) Measurement of ammonia emissions after liquid manure application: I. Construction of a wind tunnel system for measurements under field conditions. *Zeitschrift für Pflanzenernahrung und Bodenkunde* **156**, 393–396.

Brennan, B. (1993) Odour nuisance. *Water and Wastewater Treatment*, September.

Brennan, B.M., Donlon, M. and Bolton, E. (1996) Peat biofiltration as an odour control technology for sulphur based odours. *Journal of the Chartered Institution of Water and Environmental Management* **10**, 190–198.

Cain, W.S. (1980) The case against threshold measurement of emvironmental odors. *Journal of the Air Pollution Control Association* **30**, 1295–1296.

CEN (1999) *prEN 13725: Air Quality – Determination of Odour Concentration by Dynamic Olfactometry*. Draft European Standard, Brussels, Belgium.

Cheremisinoff, P.N. (1988) *Industrial Odour Control*. Oxford, Butterworth-Heinemann.

Cooper, C.D, Reinhardt, D.R, Rash, F., Seligman, D. and Keely, D. (1992) *Landfill gas emissions*. Report #92-2, Florida center for solid and hazardous waste management.

Cussler, E.L. (1997) *Diffusion, mass transfer in fluid systems*. Cambridge University Press, Cambridge.

Demmers, T.G.M, Burgess, L.R, Short, J.L., Phillips, V.R., Clark, J.A. and Wathes, C.M. (1999) Amonia emissions from two mechanically ventilated UK livestock buildings. *Atmospheric Environment* **33**, 217–227.

Denmead, O.T. (1979) Chamber system for measuring nitrous oxide emissions from soils in the field. *Soil Science Society of America Journal* **43**, 89–95.

Denmead, O.T., Simpson, J.R. and Freney, J.R. (1977) A direct field measurement of ammonia emission after injection of anhydrous ammonia. *Soil Science Society of America Journal* **41**, 1001–1004.

Dravnieks, A. and Jarke, F. (1980) Odor threshold measurement by dynamic olfactometry: significant operational variables. *Journal of the Air Pollution Control Association* **30**, 1284–1289.

Duffee, R.A. and Cha, S.S. (1980) Consideration of physical factors in dynamic olfactometry. *Journal of the Air Pollution Control Association* **30**, 1294–1295.

Eklund, B. (1999) Comparison of line- and point-source releases of tracer gases. *Atmospheric Environment* **33**, 1065–1071.

Esplin, G.J. (1988) Boundary layer emission monitoring. *Journal of the Air Pollution Control Association* **38**, 1158–1161.

Etheridge, D. and Sandberg, M. (1996) *Building ventilation: theory and measurement*. John Wiley & Sons, London.

Fan, Z.H., Zhang, J.F., Fan, C.W. and Pennise, D.M. (2001) The MMT bag for emission source sampling: Design and evaluation. *Journal of the Air and Waste Management Association* **51**(1), 60–68.

Finnigan, J. (1998) *Odour measurement and minimisation within the west of Scotland*. Paper presented at 2nd CIWEM National Conference on Odour Control in Sewage Treatment, London, 30 April 1998.

Frechen, F.B. (1992) Odour emissions of large WWTPs: source strength measurement, atmospheric dispersion calculation, emission prognosis, countermeasures – case studies. *Water Science and Technology* **25**(4–5), 375–382.

Frechen, F.B. (1994) Odour emissions of wastewater treatment plants: recent German experiences. *Water Science and Technology* **30**(4), 35–46.

Fugit, J.L., Dutaur, L., Simon, V., Riba, M.L. and Torres, L. (1996) Generation and storage of standard gas mixtures containing traces of analytes. *Fresenius Environmental Bulletin*, **5**(11–12), 682–687.

Gao, F. and Yates, S.R. (1998) Laboratory study of closed and dynamic flux chambers: experimental results and implications for field application. *Journal of Geophysical Research – Atmospheres* **103** (D20), 26115–26125.

Gostelow, P. and Parsons, S.A. (2000) Sewage treatment works odour measurement. *Water Science and Technology* **41**(6), 33–40.

Gostelow, P. and Parsons, S.A. (2001) Hydrogen sulphide measurement. In: Stuetz, R. and Frechen, F-B. (eds) *Odours in Wastewater Treatment: Measurement, Modelling and Control.* IWA Publishing, London, 120–129.

Guy, C. and Page, T. (2001) *Odour impact analysis: methodology and application?* Proceedings of the 1st IWA international conference on odour and VOCs: measurement, regulation and control techniques, The University of NSW, Sydney, Australia, 25–28 March, 237–244.

Harssema, H. and Heeres, P. (1992) A long-duration ring test of olfactometers. In: Dragt, A.J. and van Ham, J. (eds) *Biotechniques for Air Pollution Abatement and Odour Control Policies.* Elsevier Science Publishers, 395–402.

Hashmonay, R.A., Yost, M.G., Mamane, Y. and Benayahu, Y. (1999) Emission rate apportionment from fugitive sources using open-path FTIR and mathematical inversion. *Atmospheric Environment* **33**, 735–743.

Heroux, M., Gelinas, C. and Guy, C. (2001) *Odour impacts: citizen monitoring network at city of Montreal landfill site and yard waste composting facility.* Presented to 8th International Waste Management and Landfill Symposium, Sardinia (Italy), 1–5 October 2001.

Hobbs, P. (2001) Odour analysis by gas chromatography. In: Stuetz, R. and Frechen, F-B. (eds) *Odours in Wastewater Treatment: Measurement, Modelling and Control.* IWA Publishing, London, 155–178.

Hobbs, P.J, Misselbrook, T.H. and Cumby, T.R. (1999) Production and emission of odours and gases from ageing pig waste. *Journal of Agricultural Engineering Research* **72**, 291–298.

Hobbs, P.J., Misselbrook, T.H., Dhanoa, M.S. and Persaud, K.C. (2001) Development of a relationship between olfactory response and major odorants from organic wastes. *Journal of the Science of Food and Agriculture* **81**, 188–193.

Hobson, J. and Yang, G. (2001) Odour mapping using H_2S measurements. In: Stuetz, R. and Frechen, F-B. (eds) *Odours in Wastewater Treatment: Measurement, Modelling and Control.* IWA Publishing, London, 214–231.

Hodgson, B. and Paspaliaris, P. (1996) Melbourne Water's wastewater treatment lagoons: design modifications to reduce odours and enhance nutrient removal. *Water Science and Technology* **33**(7), 157–164.

Hudon, G., Guy, C. and Hermia, J. (2000) Measurement of odor intensity by an electronic nose. *Journal of the Air and Waste Management Association* **50**, 1750–1758.

Irwin, J.S. (1979) A theoretical variation of the wind profile power law exponent as a function of surface roughness and stability. *Atmospheric Environment* **13**, 191.

ISO (1994) Stationary source emissions – measurement of velocity and volume flow rate of gas streams in ducts. International Organization for Standardization, ISO 10780.

Jiang, J.K. (1996) Concentration measurement by dynamic olfactometer. *Water Environment and Technology* **8**(6), 55–58.

Jiang, J. (2001) *What are the areas for further research?* Proceedings of the 1st IWA international conference on odour and VOCs: measurement, regulation and control techniques, University of NSW, Sydney, Australia, 25–28 March, 17–26.

Jiang, K. and Kaye, R. (1996) Comparison study on portable wind tunnel system and isolation chamber for determination of VOCs from areal sources. *Water Science and Technology* **34**(3–4), 583–589.

Jiang, J.K. and Kaye, R. (1997) The selection of air velocity inside a portable wind tunnel system using odour complaint database. Control of Odors and VOC Emissions: Proceedings of Water Environment Federation Specialty Conference on Control of Odors and VOC Emissions, Houston, Texas, 20–23 April, 10–29.

Jiang, J. and Kaye, R. (2001) *Sampling techniques for odour measurement.* In: Stuetz, R. and Frechen, F-B. (eds) Odours in Wastewater Treatment: Measurement, Modelling and Control, IWA Publishing, London, 95–119.

Jiang, K., Bliss, P.J. and Schulz, T.J. (1995) The development of a sampling system for determining odor emission rates from areal sources: Part 1. Aerodynamic performance. *Journal of the Air and Waste Management Association* **45**, 917–922.

Kienbusch, M.R. (1986) *Measurement of gaseous emission rates from land surfaces using an emission flux chamber. User's guide*. **EPA/600-8/8-86/008**. US Environmental Protection Agency, Las Vegas.

Kim, D-S., Harazon, Y., Baten, M.A., Nagai, H. and Tsuruta, H. (2002) Surface flux measurements of CO_2 and N_2O from a dried rice paddy in Japan during a fallowing winter season. *Journal of the Air and Waste Management Association* **52**, 416–422.

Koe, L.C.C. (1989) Sewage odors quantification. In: Cheremisinoff, P.N. (ed.) *Encyclopedia of Environmental Control Technology. Volume 3: Wastewater Treatment Technology*, Houston, Gulf Publishing, 423–446.

Laing, D.G., Eddy, A. and Best, D.J. (1994) Perceptual characteristics of binary, trinary and quaternary odor mixtures consisting of unpleasant constituents. *Physiology and Behavior* **56**, 81–93.

Laska, M. and Hudson, R. (1991) A comparison of the detection thresholds of odour mixtures and their components. *Chemical Senses* **16**, 651–662.

Leonardos, G. (1980) Selection of panellists. *Journal of the Air Pollution Control Association* **30**, 1297.

Leyris, C., Maupetit, F., Guillot, J.M.M., Pourtier, L. and Fanlo, J.L. (2000) Laboratory study of odour emissions from areal sources: evaluation of a sampling system. *Analysis* **28**(3), 199–206.

Lindberg, S.E., Zhang, H., Vette, A.F., Gustin, M.S., Barnett, M.O. and Kuiken, T. (2002) Dynamic flux chamber measurements of gaseous mercury fluxes over soils. Part 2: Effect of flushing flow rate and verification of a two-resistance exchange interface simulation model. *Atmospheric Environment* **36**, 847–859.

Lindvall, T., Noren, O. and Thyselius, L. (1974) Odor reduction for liquid manure systems. *Transactions of the ASAE* **17**, 508–512.

Lockyer, D.R. (1984) A system for the measurement in the field of losses of ammonia through volatilisation. *Journal of the Science of Food and Agriculture* **35**, 837–848.

Loubet, B., Cellier, P., Flura, D. and Genermont, S. (1999a) An evaluation of the wind-tunnel technique for estimating ammonia volatilization from land. Part 1: Analysis and improvement for accuracy. *Journal of Agricultural Engineering Research* **72**, 71–81.

Loubet, B., Cellier, P., Flura, D. and Genermont, S. (1999b) An evaluation of the wind-tunnel technique for estimating ammonia volatilization from land. Part 2: Influence of the tunnel on transfer processes. *Journal of Agricultural Engineering Research* **72**, 83–92.

Mannheim, T., Braschkat, J. and Marschner, H. (1995) Measurement of ammonia emission after liquid manure application. II: Comparison of the wind tunnel and the IHF method under field conditions. *Zeitschrift für Pflanzenernahrung und Bodenkunde* **158**, 215–219.

McGarvey, L.J., Shorten, C.V. (2000) The effects of adsorption on the reusability of Tedlar (R) air sampling bags. *American Industrial Hygiene Association Journal* **61**(3), 375–380.

McGinn, S.M. and Janzen, H.H. (1998) Ammonia sources in agriculture and their measurement. *Canadian Journal of Soil Science* **78**, 139–148.

Miedema, H.M.E. and Ham, J.M. (1988) Odour annoyance in residential areas. *Atmospheric Environment* **22**, 2501–2507.

Miedema., H.M.E., Walpot, J.I., Vos, H. and Steunenberg, C.F. (2000) Exposure-annoyance relationships for odour from industrial sources. *Atmospheric Environment* **34**, 2927–2936.

Milton, M.J.T., Partridge, R.H. and Goody, B.A. (1995) *Minimum emission rates detectable from landfill sites using optical integrated-path techniques*. Proceedings of the A&WMA International Specialty Conference on Optical Sensing for Environmental and Process Monitoring, San Francisco, 394–404.

Milton, M.J.T, Partridge, R.H., Goody, B.A. and Andrews, A.S. (1997) *An estimate of methane emissions from UK landfill sites based on direct flux measurements at representative sites.* NPL Report QM 134.

Mosher, B.W., Czepiel, P.M., Harriss, R.C., Shorter, J.H., Kolb, C.E., McManus, J.B., Allwine, E. and Lamb, B.K. (1999) Methane emissions at nine landfill sites in the northeastern Unites States. *Environmental Science and Technology* **33**, 2088–2094.

Noble, R., Hobbs, P.J., Dobrovin-Pennington, A., Misselvrook, T.H. and Mead, A. (2001) Olfactory response to mushroom composting emissions as a function of chemical concentration. *Journal of Environmental Quality* **30**, 760–767.

Noble, R. and Hobbs, P.J. (2002) *Development of an odour management strategy for mushroom composting.* Presented at the International Conference on Odour Management and Treatment – Odour 2, 6 March 2002, Cranfield University, UK.

North, A.A. (1979) *Odours at a sewage treatment works.* Stevenage: Water Research Council Technical Report TR126.

O'Brien, M., Duffee, R.A. and Ostojic, N. (1995) Effect of sample flow rate in the determination of odor thresholds. Proceedings of the Air and Waste Management Association Conference, Bloomington MN, 13–15 September, 261–270.

Pain, B.F., Clarkson, C.R., Phillips, V.R. and Klarenbeek, J.V. (1991) Odor emission arising from application of livestock slurries on land: measurements following a micrometeorological technique and olfactometry. *Journal of Agricultural Engineering Research* **48**, 101–110.

Park, J.W. and Shin, H.C. (2001) Surface emission of landfill gas from solid waste landfill. *Atmospheric Environment* **35**, 3445–3451.

Pasquill, F. and Smith, F.B. (1983) *Atmopheric Dispersion* (3rd edn), Chichester, Ellis Horwood.

Patterson, M.Q., Stevens, J.C., Cain, W.S., and Commeto-Muniz, J.E. (1993) Detection thresholds for an olfactory mixture and its three constituent compounds. *Chemical Senses* **18**, 723–734.

Pau, J.C., Knoll, J.E. and Midgett, M.R. (1991) A tedlar bag sampling system for toxic organic-compounds in source emission sampling and analysis. *Journal of the Air and Waste Management Association* **41**, 1095–1097.

Perera, M.D.N., Hettiaratchi, J.P.A. and Achari, G. (2002) A mathematical modelling approach to improve the point estimation of landfill gas surface emissions using the flux chamber technique. *Journal of Environmental Engineering and Science* **1**, 451–462.

Pet'ka, J., Etievant, P. and Callement, G. (2000) Suitability of different plastic materials for head or nose spaces short-term storage. *Analysis* **28**(4), 330–335.

Phillips, V.R., Lane, S.J. and Burgess, L.R. (2000) A technique for measuring ammonia emissions from the individual parts of a livestock building. Proceedings of 2nd ASAE International Conference on Air Pollution from Agricultural Operations, Des Moines, Iowa, 9–11October, 84–91.

Piccot, S.D., Masemore, S.S., Ringler, E.S., Srinivasan, S., Kirchgessner, D.A. and Herget, W.F. (1994) Validation of a method for estimating pollution emission rates from area sources using open-path FTIR spectroscopy and dispersion modelling techniques. *Journal of the Air and Waste Management Association* **44**, 271–279.

Piper, A.R., Todd, L.A. and Mottus, K. (1999) A field study using open-path FTIR spectroscopy to measure and map air emissions from volume sources. *Field Analytical Chemistry and Technology* **3**(2), 69–79.

Preti, G., Gittelman, T.S., Staudte, P.B. and Luitweiler, P. (1993) Letting the nose lead the way: malodorous components in drinking water. *Analytical Chemistry* 65, 699A–702A.

Reitz, P. and Kutzbach, H.D. (1997) *Accuracy of a wind tunnel system for measuring ammonia emissions after slurry application.* Proceedings of the conference on Ammonia and Odour Emissions from Animal Production Facilities, 6–10 October, Vinkeloord, The Netherlands, Vol II, 595–602.

Rix, M.R.E. (2001) *Odour sample pre-dilution: when and how much?* Proceedings of the 1st IWA international conference on odour and VOCs: measurement, regulation and control techniques, The University of NSW, Sydney, Australia, 25–28 March, 155–161.

Sadek, S.E., Lee, T.C., Smith, J.D. and Gebel, R. (1998) Flux chamber measurements of mass transfer rates. *Journal of Environmental Engineering* **124**, 111–121.

Schmidt, D.R. and Bicudo, J.R. (2000) *Odor and hydrogen sulfide emissions during the composting of caged layer manure*. Proceedings of 2nd ASAE International Conference on Air Pollution from Agricultural Operations, Des Moines, Iowa, 9–11 October, 75–83.

Schuetzle, D., Prater, T.J. and Ruddell, S.R. (1975) Sampling and analysis of emissions from stationery sources. I: Odor and total hydrocarbons. *Journal of the Air Pollution Control Association* **25**, 925–932.

Schulz, T.J. and van Harreveld, A.P. (1996) International moves towards standardisation of odour measurement using olfactometry. *Water Science and Technology*, **34**(3/4), 541–547.

Simpson, O.A. and Kagan, R.H. (1990) *Measurements of emissions at a chemical waste water site with an open path remote fourier transform interferometer*. Proceedings of the 1990 EPA/AWMA International Symposium on the Measurement of Toxic and Related Air Pollution, Raleigh, North Carolina, 937–939.

Smith, R.J. (1995) A Gaussian model for estimating odour emissions from area sources. *Mathematical Computer Modelling* **21** (9), 23–29.

Smith, N. (1998) H_2S dispersion modelling. MSc Thesis, Cranfield University.

Smith, R.J. and Watts, P.J. (1994a) Determination of odour emission rates from cattle feedlots. Part I: A review. *Journal of Agricultural Engineering Research* **57**, 145–155.

Smith, R.J. and Watts, P.J. (1994b) Determination of odour emission rates from cattle feedlots. Part II: Evaluation of two wind tunnels of different size. *Journal of Agricultural Engineering Research* **58**, 231–240.

Sneath, R.W. (2001) *Olfactometry and the CEN standard prEN 17325*. In: Stuetz, R. and Frechen, F-B. (eds) *Odours in Wastewater Treatment: Measurement, Modelling and Control*, IWA Publishing, London, 95–119.

Sneath, R.W. and Clarkson, C. (2000) Odour measurement: a code of practice. *Water Science and Technology* **41**(6), 25–31.

Stern, L., Baisden, W.T. and Amundson, R. (1999) Processes controlling the oxygen isotope ratio of soil CO_2: analytic and numerical modeling. *Geochimica et Cosmochimica Acta* **63**, 799–814.

Stordeur, R.T., Stordeur, C.M., Levine, S.P. and Hoggatt, J.H. (1981) A new microprocessor-controlled dynamic olfactometer. *Journal of the Air Pollution Control Association* **31**, 377–380.

Sulyok, M., Haberhauer-Troyer, C., Rosenberg, E. and Grasserbauer, M. (2001) Investigation of the storage stability of selected volatile sulfur compounds in different sampling containers. *Journal of Chromatography A* **917**, 367–374.

Sweeten, J.M., McFarland, A.R., Sorel, J.E., Gauntt, R.O. and Reddell, D.L. (1988) A butanol olfactometer for field measurement of ambient odors. *Journal of the Air Pollution Control Association* **34**, 1208–1213.

Thibodeaux, L.J. (1996) *Environmental Chemodynamics* (2nd edn). John Wiley and Sons, New York.

Tillman, F.D., Choi, J-W. and Smith, J.A. (2002) *Design and testing of a chamber device to measure organic vapor fluxes from the unsaturated zone under natural conditions*. Presented to CSCE/EWRI of ASCE Environmental Engineering Conference, Niagra, 2002.

Todd, L.A., Ramanthan, M., Mottuse, K., Katz, R., Dodson, A. and Mihlan, G. (2001) Measuring chemical emissions using open-path fourier transform infrared (OP-FTIR) spectroscopy and computer-assisted tomography. *Atmospheric Environment* **35**, 1937–1947.

van Belois, H.J. and Anzion, C.J.M. (1992) *Measurement of emissions over surface areas using the hood method*. In: Dragt, A.J. and van Ham, J. (eds) *Biotechniques for Air Pollution Abatement*, Elsevier Science Publishers.

van der Weerden, T.J., Moal, J.F., Martinez, J., Pain, B.F. and Guiziou, F. (1996) Evaluation of the wind-tunnel method for measurement of ammonia volatilization from land. *Journal of Agricultural Engineering Research* **64**, 11–14.

van Harreveld, A., Heeres, P. and Harssema, H. (1999) A review of 20 years of standardization of odour concentration measurement by dynamic olfactometry in Europe. *Journal of the Air and Waste Management Association* **49**(6), 705–715.

van Langenhove, H. and van Broeck, G. (2001) Applicability of sniffing team observations: experience of field measurements. *Water Science and Technology* 44(9), 65–70.

Vincent, A. and Hobson, J. (1998) *Odour Control*. CIWEM Monographs on Best Practice No. 2, London: Chartered Institution of Water and Environmental Management.

Walker, J.C. (2001) *The performance of the human nose in odour measurement.* Proceedings of the 1st IWA international conference on odour and VOCs: measurement, regulation and control techniques, The University of NSW, Sydney, Australia, 25–28 March, 1–7.

Wang, X. and Kaye, R. (2001) *Improvement of wind tunnel sampling system for odour and VOC.* Proceedings of the 1st IWA international conference on odour and VOCs: measurement, regulation and control techniques, The University of NSW, Sydney, Australia, 25–28 March, 147–154.

Wang, Y., Raihala, T.S., Jackman, A.P. and StJohn, R. (1996) Use of Tedlar bags in VOC testing and storage: evidence of significant VOC losses. *Environmental Science and Technology* **30**(10), 3115–3117.

Whitcraft, W.K. and Wood, K.N. (1990) *Use of remote sensing to measure wastewater treatment plant emissions*. Proceedings of the 83rd Annual Meeting and Exhibition. Pittsburgh, Pennsylvania.

Wilson, J.D., Thurtell, G.W., Kidd, G.E. and Beauchamp, E.G. (1982) Estimation of the rate of gaseous mass transfer from a surface plot to the atmosphere. *Atmospheric Environment* **16**, 1861–1867.

Wilson, J.D., Catchpoole, V.R., Denmead, O.T. and Thurtell, G.W. (1983) Verification of a simple micrometeorological method for estimating the rate of gaseous mass transfer from the ground to the atmosphere. *Agricultural Meteorology* **29**, 183–189.

Winegar, E.D. and Schmidt, C.C. (1998) Jerome 631-X portable hydrogen sulphide sensor: laboratory and field evaluation. Report to Arizona Instrument Corporation.

Wright, R.H. (1982) *The sense of smell*. CRC Press, Florida.

Young, P.J. (1984) Odours from effluent and waste treatment. *Effluent and Water Treatment Journal* **24**, 189–195.

Zhang, H., Lindberg, S.E., Barnett, M.O., Vette, A.F. and Gustin, M.S. (2002) Dynamic flux chamber measurement of gaseous mercury emission fluxes over soils. Part 1: simulation of gaseous mercury emissions from soils using a two-resistance exchange interface model. *Atmospheric Environment* **36**, 835–846.

Appendix A: Odorants associated with waste treatment

Class	Compound	Formula	Character	Threshold (ppb)
Sulphurous	Hydrogen sulphide	H_2S	Rotten eggs	0.5–1.1
	Methyl mercaptan	CH_3SH	Decayed cabbage, garlic	0.0014–18
	Ethyl mercaptan	CH_3CH_2SH	Decayed cabbage	0.019–0.32
	Allyl mercaptan	CH_2CHCH_2SH	Garlic	0.05
	Phenyl mercaptan	C6H5SH	Putrid, decay	0.062
	Benzyl mercaptan	$C_6H_5CH_2SH$	Unpleasant	0.19
	Crotyl mercaptan	$CH_3CHCHCH_2SH$	Skunk, rancid	0.029
	Butyl mercaptan		Unpleasant	1
	tButyl mercaptan	$(CH_3)_3CSH$	Unpleasant	0.08
	Propyl mercaptan		Unpleasant	0.5
	Dimethyl sulphide	$(CH_3)_2S$	Decayed vegetables, garlic	0.12–10
	Diethyl sulphide	$(C_2H_5)_2S$	Nauseating, ether	0.3
	Diphenyl sulphide	$(C_6H_5)_2S$	Unpleasant, burnt rubber	0.048
	Diallyl sulphide		Garlic	0.14
	Dimethyl disulphide	$(CH_3)_2S_2$	Putrification	0.3–11
	Carbon disulphide	CS_2	Decayed vegetables	210
	Thiocresol	$CH_3C_6H_4SH$	Skunk, rancid	0.1
	Sulphur dioxide	SO_2	Sharp, pungent, irritating	9

© IWA Publishing. *Sampling for the Measurement of Odours*. P. Gostelow, P. Longhurst, S. A. Parsons and R. M. Stuetz. ISBN: 1 84339 033 7

Class	Compound	Formula	Character	Threshold (ppb)
Nitrogenous	Ammonia	NH_3	Sharp, pungent	37–15300
	Methylamine	CH_3NH_2	Fishy	0.9–53
	Dimethylamine	$(CH_3)_2NH$	Fishy	23–80
	Trimethylamine	$(CH_3)_3N$	Fishy, ammoniacal	0.2
	Ethylamine	$C_2H_5NH_2$	Ammoniacal	2400
	Diethylamine	$(C_2H_5)_2NH_2$	Fishy, ammoniacal	0.02–14
	Triethylamine	$(C_2H_5)_3N$	Fishy, ammoniacal	0.1–0.65
	Diamines, i.e. Cadaverine	$NH_2(CH_2)_5NH_2$	Decomposing meat	–
	Pyridine	C_6H_5N	Disagreeable, irritating	3.7
	Indole	C_8H_6NH	Faecal, nauseating	1.4
	Scatole or Skatole	C_9H_8NH	Faecal, nauseating	0.002–1.2
Acids	Acetic (ethanoic)	CH_3COOH	Vinegar	16
	Butyric (butanoic)	C_3H_7COOH	Rancid, sweaty	0.09–20
	Valeric	C_4H_9COOH	Sweaty	0.6–2630
	Ethyl acrylate	$CH_2CHCOOC_2H_5$	Earthy	–
Aldehydes and ketones	Formaldehyde	$HCHO$	Acrid, suffocating	370
	Acetaldehyde	CH_3CHO	Fruit, apple	0.005–66
	Butyraldehyde	C_3H_7CHO	Rancid, sweaty	4.6
	Isobutyraldehyde	$(CH_3)_2CHCHO$	Fruit	4.7–7
	Isovaleraldehyde	$(CH_3)_2CHCH_2CHO$	Fruit, apple	0.7–9
	Acetone	CH_3COCH_3	Fruit, sweet	4580
	Butanone	$C_2H_5COCH_3$	Green apple	270

Appendix B: Terms and definitions used in the draft CEN standard prEN 13725 'Odour Concentration Measurement by Dynamic Olfactometry'

accepted reference value: A value that serves as an agreed-upon reference for comparison, and which is derived as a consensus value, based on collaborative experimental work under the auspices of a scientific or engineering group. [ISO 5725 part 1, abridged]

accuracy: Closeness of agreement between test result and the accepted reference value. [ISO 5725 part 1]. **Note:** The term 'accuracy', when applied to a set of observed values, describes a combination of random components and a common systematic error or bias component.

(sensory) adaptation: Temporary modification of the sensitivity of a sense organ due to continued and/or repeated stimulation. [ISO 5492:1992]

anosmia: Lack of sensitivity to olfactory stimuli. [ISO 5492:1992]

assessor: Somebody who participates in odour testing.

bias: The difference between the expectation of the test results and an accepted reference value. [ISO 5725 part 1]. **Note**: Bias is often called 'systematic error'

certified reference material, CRM: A reference material of which one or more property values are certified by a technically valid procedure accompanied by or traceable to a certificate or other documentation which is issued by a certifying body. [ISO 5725 part 4]. **Note**: For the purpose of olfactometry the Netherlands Measuring Institute in Delft certifies reference materials by comparing them with national standard gas mixtures. A European certifying body for gases does not currently exist.

delayed olfactometry: Measurement of an odour with a time-lag between sampling and measurement. The odour sample is preserved in an appropriate container. [AFNOR X43-104E, see bibliography, Annex J.]

detection limit: see 'lower detection limit'.

detection threshold, (for a reference material)**:** The odorant concentration which has a probability of 0.5 of being detected under the conditions of the test.

detection threshold (for an environmental sample)**:** The dilution factor at which the sample has a probability of 0.5 of being detected under the conditions of the test.

diffuse sources: Sources with defined dimensions (mostly surface sources) which do not have a defined waste airflow, such as waste dumps, lagoons, fields after manure spreading, un-aerated compost piles.

dilution factor: The dilution factor is the ratio between flow or volume after dilution and the flow or volume of the odorous gas. [AFNOR X 43-104E, see bibliography, Annex J]

dilution series: The presentation of a sequence of dilutions to one panel member in order to obtain one Individual Threshold Estimate. **Note**: One dilution series can consist of:

– One series of presentations, at odour concentrations in ascending or random order, where, when sorted in order of descending concentrations, a significant change from consistently TRUE responses to a FALSE response occurs.

– A repeated pattern of presentations according to the procedure described for forced choice/probability mode.

direct olfactometry: Measurement of odour concentrations without any time-lag between the sampling (operation) and the measurements; equivalent to dynamic sampling or on-line olfactometry. [AFNOR X 43-104E, see bibliography, Annex J].

dynamic olfactometer: A dynamic olfactometer delivers a flow of mixtures of odorous and neutral gas with known dilution factors in a common outlet. [AFNOR X 43-101E, modified, see bibliography, Annex J]

dynamic olfactometry: Olfactometry using a dynamic olfactometer.

dynamic sampling: Sampling in direct olfactometry.

European Odour unit: That amount of odorant(s) that, when evaporated into 1 cubic metre of neutral gas at standard conditions, elicits a physiological response from a panel (detection threshold) equivalent to that elicited by one European Reference Odour Mass (EROM), evaporated in one cubic metre of neutral gas at standard conditions

European Reference Odour Mass, EROM: The accepted reference value for the European odour unit, equal to a defined mass of a certified reference material.

One EROM is equivalent to 123 μg n-butanol (CAS 71-36-3). Evaporated in 1 cubic metre of neutral gas this produces a concentration of 0,040 μmol/mol.

expected value: The value approached by the average value with an increasing number of measurement values.

forced choice method: For this standard the following definition applies: an olfactometric method in which assessors are forced to make a choice out of two or more airflows, one of which is the diluted sample, even if no difference is observed.

fugitive sources: Elusive or difficult to identify sources releasing undefined quantities of odorants, e.g. valve and flange leakage, passive ventilation apertures etc.

group threshold: Detection threshold applying to a group of assessors.

identification threshold: See 'recognition threshold'.

individual threshold: Detection threshold applying to an individual.

individual threshold estimate, ITE : The detection threshold applying to an individual estimated on the basis of one dilution series.

instability: The change of a characteristic over a stated period of time, consisting of a systematic part (drift) and a random part (dispersion). [ISO 9169, 6.2.2.]

instrumental dilution range: Range between the minimum and maximum dilution factor.

instrumental fall time: Time taken for the reading to pass from (by convention) 10% to (by convention) 90% of the final change in output signal reading. [ISO 6879]

For instruments where transient oscillations occur in the approach to the final output signal reading, the fall time is the time taken for the instrument reading to pass from (by convention) 10% of the final change in instrument reading until the oscillations fall to less than 10% (by convention) of the final change in instrument reading.

instrumental lag time: Time taken to reach 10 % (by convention) of the final change in instrument reading. [ISO 6879]

instrumental response time: Time taken for an instrument to respond to an abrupt change in value of the air quality characteristic. It is the sum of the lag time and rise time (rising mode) or lag time and fall time (falling mode). [ISO 6879, modified]

instrumental rise time: Time taken for the reading to pass from (by convention) 10% to (by convention) 90% of the final change in output signal reading. [ISO 6879]

For instruments where transient oscillations occur in the approach to the final output signal reading, the rise time is the time taken for the instrument reading to pass from (by convention) 10% of the final change in instrument reading until the oscillations fall to less than (by convention) 10% of the final change in instrument reading.

lower detection limit, LDL: Lowest value of the air quality characteristic which, with 95% probability, can be distinguished from a zero sample. [ISO 6879]

maximum dilution factor: Maximum settable dilution factor of the olfactometer; an instrument property.

measurement: The presentation to all panel members of those dilution series necessary to produce sufficient data to calculate the odour concentration for one sample.

measuring range: The measuring range comprises all odour concentrations which can be measured by a specific olfactometer. It depends on the minimum and maximum dilution factor and the step factor. The numerical values defining the measuring range are the minimum dilution factor multiplied with the step factor to the power three and the maximum dilution factor divided by the step factor to the power three.

minimum dilution factor: Minimum settable dilution factor of the olfactometer; an instrument property.

neutral gas: Air or nitrogen that is treated in such a way that it is as odourless as possible and that does, according to panel members, not interfere with the odour under investigation. **Safety Warning**: Nitrogen is only used to predilute the sample itself. For the olfactometer the neutral gas used to dilute the sample and present a reference shall be air.

objective method: Any method in which the effects of personal opinions are minimised. [ISO 5492]

odorant: A substance which stimulates a human olfactory system so that an odour is perceived. [Hangartner, M, *et al.* 1989, see bibliography, Annex J]

odorant flow rate: The odorant flow rate is the quantity of odorous substances passing through a defined area at each time unit. It is the product of the odour concentration c_{od}, the outlet velocity v and the outlet area A or the product of the odour concentration c_{od} and the pertinent volume flow rate $\dot{V}$. Its unit is ou_E/h (or ou_E/min or ou_E/s, respectively). **Note**: The odorant (emission) flow rate is the quantity equivalent to the emission mass or volume flow rate, for example in dispersion models.

odorous gas: Gas that contains odorants.

odour: Organoleptic attribute perceptible by the olfactory organ on sniffing certain volatile substances. [ISO 5492]

odour abatement efficiency: The reduction of the odour concentration or the odorant flow rate due to an abatement technique, expressed as a fraction (or percentage) of the odour concentration in or the odorant flow rate of the untreated gas stream.

odour concentration: The number of European odour units in a cubic metre of gas at standard conditions. **Note**: The odour concentration is not a linear measure for the intensity of an odour. Steven's Law describes the a-linear relation between odour stimulus and its perceived intensity. When using odour concentrations in dispersion modelling, the issue is complicated by the effects of the averaging time of the dispersion model, further complicating the use of the odour concentration as a direct measure for dose. To define a 'no nuisance level', the entire method of dosage evaluation, including the dispersion model, will yield a 'dose'. The relation between this 'dose' and its effect (odour annoyance) should be validated in practical situations to be a useful predictive tool for occurrence of odour nuisance.

odour detection: To become aware of the sensation resulting from adequate stimulation of the olfactory system.

odour panel: See 'panel'.

odour unit: One odour unit is the amount of (a mixture of) odorants present in one cubic metre of odorous gas (under standard conditions) at the panel threshold.
Note: See also 'European odour unit'.

odour threshold: See 'panel threshold'.

odourless gas : See 'neutral gas'.

olfactometer: Apparatus in which a sample of odorous gas is diluted with neutral gas in a defined ratio and presented to assessors.

olfactometry: Measurement of the response of assessors to olfactory stimuli. [ISO 5492]

olfactory: Pertaining to the sense of smell. [ISO 5492]

olfactory receptor: Specific part of the olfactory system which responds to an odorant. [after ISO 5492]

olfactory stimulus: That which can excite an olfactory receptor. [ISO 5492, modified]

on-line olfactometry: See 'direct olfactometry'.

operator: Person directly involved in operating the olfactometer and instructing the panel in olfactometry.

panel: A group of panel members.

panel member: An assessor who is qualified to judge samples of odorous gas, using dynamic olfactometry within the scope of this standard.

panel screening: Procedure to determine if the performance of panel members is in compliance with selection criteria. See also 'panel selection'.

panel selection: Procedure to determine which assessors are qualified as panel members.

panel threshold: Detection threshold applying to a panel.

perception: Awareness of the effects of single or multiple sensory stimuli. [ISO 5492]

population (detection) threshold: Detection threshold applying to the general population, if this population is not specified.

precision: Closeness of agreement between independent test results obtained under prescribed conditions. [ISO 5725-part 1]. **Note**: Precision depends only on the distribution of random errors and does not relate to the true value or the accepted reference value. The measure of precision is usually expressed in terms of imprecision and computed as a standard deviation of the test results. Higher imprecision is reflected by a larger standard deviation. 'Independent test results' means results obtained in a manner not influenced by any previous result on the same or similar material.

presentation: One presentation is the presentation of one dilution to one assessor. [NVN 2820]

presentation series: The presentation of one dilution to all panel members in one round.

presented gas flow: The gas flow presented to the assessor. It may be:

– a diluted odour sample

– neutral gas (e.g. as a blank or reference air).

proficiency testing: The system for objectively testing laboratory results by an external agency.

quality: The totality of features and characteristics of a product or service that bear on its ability to satisfy stated or implied needs. [ISO 6879]

quality assurance: All those planned and systematic actions necessary to provide adequate confidence that a product, process or service will satisfy given requirements for quality. [ISO 6879]

random error: Unpredictable errors which average to zero. [ISO 5492]

recognition threshold: The odour concentration which has a probability of 0.5 of being recognised under the conditions of the test (definition not applied in this standard).

reference material: For this International Standard: Substance or mixture of substances, the composition of which is known within specified limits, and one or more of the properties of which is sufficiently well established to be used for the calibration of an apparatus, the assessment of a measuring method, or for assigning values to materials. [ISO 6879]

reference value: See 'accepted reference value'.

repeatability: Precision under repeatability conditions. [ISO 5725-part 1]

repeatability conditions: Conditions where independent test results are obtained with the same method on identical test material in the same laboratory by the same operator using the same equipment within short intervals of time. [ISO 5725-part 1]

repeatability limit: The value less than or equal to which the absolute difference between two test results obtained under repeatability conditions may be expected to be with a probability of 0,95. [ISO 5725-part 1, modified]. **Note**: In this standard the test result is the decimal logarithm of the panel threshold.

reproducibility: Precision under reproducibility conditions. [ISO 5725-part 1]

reproducibility conditions: Conditions where test results are obtained with the same method on identical test material in different laboratories with different operators using different equipment. [ISO 5725-part 1]

reproducibility limit: The value less than or equal to which absolute difference between two test results obtained under reproducibility conditions may be expected to be with a probability of 0,95. [ISO 5725-part 1, modified]. **Note**: In this standard the test result is the decimal logarithm of the panel threshold.

responsible person: Person who is finally responsible for the total of olfactometry in a laboratory.

round: One round is the presentation of one dilution series to all assessors.

sample: In the context of this standard, the sample is the odorous gas sample. It is an amount of gas which is assumed to be representative of the gas mass or gas flow under investigation, and which is examined for odour concentration. [ISO 6879]

sensory fatigue: Form of adaptation in which a decrease in sensitivity occurs. [ISO 5492]

sensory reference: The presented gas flow to which the diluted sample is compared.

single measurement: Identical to 'measurement', see also 'test result'.

to smell: To detect or to attempt to detect an odorant.

standard conditions for olfactometry: At room temperature (293 K), normal atmospheric pressure (101.3 kPa) on a wet basis [as in ISO 10780].

Note: This applies both to olfactometric measurements and volume flow rates of emissions.

static olfactometer: A static olfactometer dilutes by mixing two known volumes of gas, odorous and odourless, respectively. The rate of dilution is calculated from the volumes. [AFNOR X 43-101E]

static sampling: Sampling in delayed olfactometry.

step factor: The factor by which each dilution factor in a dilution series differs from adjacent dilutions.

subjective method: Any method in which the personal opinions are taken into consideration. [ISO 5492]

substance: Species of matter of definite chemical composition. [Hangartner, 1989]

test result: The value of a characteristic obtained by completely carrying out a specific measurement, once.[ISO 5725-part 1]

traceability: The property of the result of a measurement that can be related through an unbroken chain of comparisons to appropriate reference materials, generally national or international reference materials, using measurement standards of successively increasing accuracy.

trueness: The closeness of agreement between the average value obtained from a large series of test results and an accepted reference value. [ISO 5725-part 1]. **Note**: The measure of trueness usually is expressed in terms of bias.

true value: See 'accepted reference value'.

yes/no method: Olfactometric method in which assessors are asked to judge whether an odour is detected or not.

Index

IWA Publishing's authorised EU representative for General Product Safety Regulations is Diane D'Arras, 15 rue Duret, 75116 Paris, France, e-mail: safety@iwap.co.uk.

Printed and bound by CPI Group (UK) Ltd, Croydon, CR0 4YY
13/05/2026
02109606-0003